Texts and
Monographs
in Physics

W. Beiglböck
series editor

Polarized Electrons

Joachim Kessler

Springer-Verlag
Berlin Heidelberg New York
1976

Professor Dr. JOACHIM KESSLER

Physikalisches Institut der Universität
Schlossplatz 7, D-4400 Münster

Professor Dr. WOLF BEIGLBÖCK

Institut für Angewandte Mathematik der Universität
Im Neuenheimer Feld 294, D-6900 Heidelberg

With 104 figures

ISBN 3-540-07678-6 Springer-Verlag Berlin Heidelberg New York
ISBN 0-387-07678-6 Springer-Verlag New York Heidelberg Berlin

Library of Congress Cataloging in Publication Data. Kessler, Joachim, 1930–. Polarized electrons. (Texts and monographs in physics). Bibliography: p. Includes index. I. Electrons—Polarization. I. Title. QC793.5.E628K47 539.7'2112 76–9863

This work is subject to copyright. All rights are reserved, whether the whole or part of the material is concerned, specifically those of translation, reprinting, re-use of illustrations, broadcasting, reproduction by photocopying machine or similar means, and storage in data banks. Under § 54 of the German Copyright Law, where copies are made for other than private use, a fee is payable to the publisher, the amount of the fee to be determined by agreement with the publisher.

© by Springer-Verlag Berlin Heidelberg 1976
Printed in Germany

The use of registered names, trademarks, etc. in this publication does not imply, even in the absence of a specific statement, that such names are exempt from the relevant protective laws and regulations and therefore free for general use.

Offset printing and bookbinding: Konrad Triltsch, Graphischer Betrieb, Würzburg

Preface

This book deals with the physics of spin-polarized free electrons. Many aspects of this rapidly expanding field have been treated in review articles, but to date a self-contained monograph has not been available.

In writing this book, I have tried to oppose the current trend in science that sees specialists writing primarily for like-minded specialists, and even physicists in closely related fields understanding each other less than they are inclined to admit. I have attempted to treat a modern field of physics in a style similar to that of a textbook.

The presentation should be intelligible to readers at the graduate level, and while it may demand concentration, I hope it will not require deciphering. If the reader feels that it occasionally dwells upon rather elementary topics, he should remember that this pedestrian excursion is meant to be reasonably self-contained. It was, for example, necessary to give a simple introduction to the Dirac theory in order to have a basis for the discussion of Mott scattering—one of the most important techniques in polarized-electron studies.

This monograph is intended to be an introduction to the field of polarized electrons and not a replacement for review articles on the individual topics discussed. It does not include electron polarization in β decay, a field which has been covered in other books. Areas such as electron spin resonance, in which it is not the spins of *free* electrons that are oriented, are beyond the scope of this book. Well-established areas, like Mott scattering, have naturally been treated in more detail than areas that are just starting to develop, such as high-energy electron scattering. Ideas or general results that have not been quantitatively established, theoretically or experimentally, have not been considered, since physical results must be put on a quantitative basis.

Following two introductory chapters, the polarization effects in electron scattering from unpolarized targets are discussed in Chapter 3. Processes governed by electron exchange are treated in Chapter 4. The numerous spin-polarization effects in ionization processes are discussed in Chapter 5, and the following chapter deals with electron polarization in solid-state physics. At a time when scientists are frequently asked about the practical benefits of their work, it seems particularly appropriate to consider the applications of polarized-electron physics. This is done in Chapter 7 which also includes a section on the future prospects in this field.

In keeping with the introductory character of the book, the main purpose of the reference lists is to aid the reader in completing or supplementing the information in certain sections. The newcomer to the field should refer to the review articles wherever they exist. Primary sources have been cited if they are directly referred to in the text or if they have not yet been listed in review papers or other references.

It is a pleasure to express my gratitude to the many people who have contributed to the completion of this project. Several sections have been considerably influenced by the ideas and achievements of my coworkers— particularly Drs. G. F. HANNE, U. HEINZMANN, and K. JOST— with whom, over the years, I have studied many of the topics discussed. The generous hospitality of the Joint Institute for Laboratory Astrophysics gave me the chance to write this book. The stimulating atmosphere of JILA which I enjoyed during my stay as a Visiting Fellow provided the ideal setting for this project. I gratefully acknowledge the excellent work of the JILA editorial office; thanks to the numerous helpful suggestions from L. VOLSKY and the typing skill of G. ROMEY, the transformation (in record time) of my stacks of messy, marked-up sheets into a beautiful manuscript was a joy to behold. I am particularly grateful to Dr. M. LAMBROPOULOS who was kind enough to read the entire manuscript; her constructive criticisms have improved it considerably. Discussions with Prof. H. MERZ in Münster and with many colleagues in Boulder have helped to clarify several passages. I appreciate the application and conscientiousness of H. GERBERON and B. GÖHLSDORF who prepared most of the illustrations. I am also grateful for the assistance of E. RUSSEL and Dr. C. B. LUCAS in translating a number of my lectures which I used in preparing parts of the manuscript. Finally, I wish to thank those listeners, at home and abroad, who, by their reactions to my lectures, have helped to clarify this presentation.

Boulder, Colorado
August, 1975

JOACHIM KESSLER

Contents

1. Introduction

1.1 The Concept of Polarized Electrons 1
1.2 Why Conventional Polarization Filters Do Not Work with Electrons 2

2. Description of Polarized Electrons

2.1 A Few Results from Elementary Quantum Mechanics . . . 7
2.2 Pure Spin States 9
2.3 Statistical Mixtures of Spin States. Description of Electron Polarization by Density Matrices 14

3. Polarization Effects in Electron Scattering from Unpolarized Targets

3.1 The Dirac Equation and Its Interpretation 21
3.2 Calculation of the Differential Scattering Cross Section . . . 33
3.3 The Role of Spin Polarization in Scattering 40
    3.3.1 Polarization Dependence of the Cross Section 40
    3.3.2 Polarization of an Electron Beam by Scattering . . . 44
    3.3.3 Behavior of the Polarization in Scattering 45
    3.3.4 Double Scattering Experiments 49
3.4 Simple Physical Description of the Polarization Phenomena 51
    3.4.1 Illustration of the Rotation of the Polarization Vector 52
    3.4.2 Illustration of the Change in the Magnitude of the Polarization Vector 52
    3.4.3 Illustration of the Asymmetry in the Scattering of a Polarized Beam 55
    3.4.4 Transversality of the Polarization as a Consequence of Parity Conservation. Counterexample: Longitudinal Polarization in β Decay 56
    3.4.5 Equality of Polarizing and Analyzing Power 59

3.5 Quantitative Results 61
  3.5.1 Coulomb Field 61
  3.5.2 Screened Coulomb Field 63
3.6 Experimental Investigations 68
  3.6.1 Double Scattering Experiments 69
  3.6.2 Triple Scattering Experiments 72
  3.6.3 Experimental Equipment: Mott Detectors and
        Polarization Transformers 76
3.7 Inelastic Scattering, Resonance Scattering, Electron-Molecule
    Scattering. Further Processes Used for Polarization Analysis 83

4. Exchange Processes in Electron-Atom Scattering

4.1 Polarization Effects in Elastic Exchange Scattering 87
4.2 Experiments on Polarization Effects in Elastic Exchange
    Scattering 95
4.3 Polarization Effects in Inelastic Exchange Scattering 100
  4.3.1 One-Electron Atoms 100
  4.3.2 Two-Electron Atoms 109
4.4 Møller Scattering 116

5. Polarized Electrons by Ionization

5.1 Photoionization of Polarized Atoms 123
5.2 Fano Effect 125
  5.2.1 Theory of the Fano Effect 125
  5.2.2 Illustration of the Fano Effect. Experimental Results 133
5.3 Autoionizing Transitions 139
5.4 Multiphoton Ionization 143
5.5 Collisional Ionization of Polarized Atoms 147
  5.5.1 Collisional Ionization of Polarized Metastable
        Deuterium Atoms 147
  5.5.2 Penning Ionization 149

6. Polarized Electrons from Solids

6.1 Magnetic Materials 153
  6.1.1 Photoemission 155
  6.1.2 Field Emission 159
6.2 Nonmagnetic Materials 163

7. Further Applications and Prospects

7.1 Investigations of the Structure of Matter 171
    7.1.1 Low-Energy Electron Diffraction (LEED) 171
    7.1.2 Electron-Molecule Scattering 179
    7.1.3 Electron Microscopy 181
    7.1.4 Why Isn't Nature Ambidextrous? 182
    7.1.5 High-Energy Physics 183
7.2 $g-2$ Experiments for Measuring the Anomalous Magnetic Moment of the Electron. Electron Maser 185
7.3 Sources of Polarized Electrons 195
7.4 Prospects 205

References 211

Subject Index 217

1. Introduction

1.1 The Concept of Polarized Electrons

An ensemble of electrons is said to be polarized if the electron spins have a preferential orientation so that there exists a direction for which the two possible spin states are not equally populated. Reasons are given for the interest in polarized electrons.

In early experiments with free electrons the direction of their spins was seldom considered. The spins in electron beams that were produced by conventional methods (such as thermal emission or the photoelectric effect) had arbitrary directions. Whenever the spin direction played a role, one had to average over all spin orientations in order to describe the experiments properly.

Only in recent years has it been found possible to produce electron beams in which the spins have a preferential orientation. They are called polarized electron beams in analogy to polarized light in which it is the field vectors that have a preferred orientation. To put it more precisely: An electron beam (or any other electron ensemble) is said to be polarized if there exists a direction for which the two possible spin states are not equally populated.

If all spins have the same direction one has the extreme case of a totally polarized ensemble of electrons (Fig. 1.1). If not all, but only a majority of the spins has the same direction, the ensemble is called partially polarized.

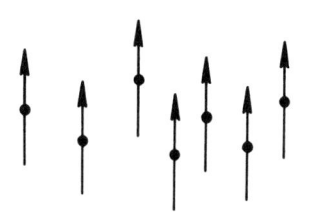

Fig. 1.1. Ensemble of totally polarized electrons

There are many reasons for the interest in polarized electrons. One essential reason is that in physical investigations one endeavors to define as exactly as possible the initial and/or final states of the systems being considered. Let us illustrate this statement with two examples. It is important in many electron-scattering experiments to be able to select electrons of as uniform energy as possible. Otherwise one would have to carry out complicated averaging in order to understand the results, and many experiments (e.g., observing the excitation of particular energy states of atoms) could not even be performed. This also applies to momentum: Often one endeavors to have electrons in the form of a well-defined beam, that is, a beam in which the directions of the momenta of the individual electrons are as uniform as possible. A swarm of electrons with arbitrary momentum directions would, for example, be unsuitable for bombarding a target. For quite analogous reasons, it is important in the investigation of the large number of spin-dependent processes that occur in physics to have electrons available in well-defined spin states. Thus one is not obliged to average over all possibilities that may arise from different spin directions, thereby losing valuable information. One can rather investigate the individual possibilities separately.

This somewhat general statement will be substantiated in later chapters. Numerous other reasons for investigations with polarized electrons will then become clear, e.g., for obtaining a better understanding of the structure of magnetic substances or for determining precisely the magnetic moment of the electron.

1.2 Why Conventional Polarization Filters Do Not Work with Electrons

Conventional spin filters, the prototype of which is the Stern-Gerlach magnet, do not work with free electrons. This is because a Lorentz force which does not appear with neutral atoms arises in the Stern-Gerlach magnet. This, combined with the uncertainty principle, prevents the separation of spin-up and spin-down electrons.

When MALUS in 1808 looked through a calcite crystal at the light reflected from a windowpane of the Palais Luxembourg, he detected the polarization of light. When STERN and GERLACH in 1921 sent an atomic beam through an inhomogeneous magnetic field they detected the polarization of atoms. Numerous exciting experiments with polarized light or polarized atoms have been made since these early discoveries. However, experiments of comparable quality with polarized electrons have been possible only in the past one or two decades.

1.2 Why Conventional Polarization Filters Do Not Work with Electrons

This is not accidental; the reason can be easily given. Polarized light can be produced from unpolarized light by sending it through a polarizer which eliminates one of the two basic directions of polarization. One therefore loses a factor of 2 in intensity. Similarly, a polarized atomic beam can be produced by sending an unpolarized atomic beam through a spin filter. If, for example, an alkali atomic beam passes through a Stern-Gerlach magnet, it splits into two beams with opposite spin directions of the valence electrons. One can eliminate one of these beams and thus again have a polarized beam with an intensity loss of a factor of 2.

This procedure does not work with electrons. It is fundamentally impossible to polarize free electrons with the use of a Stern-Gerlach experiment as can be seen in the following [1.1].

In Fig. 1.2 the electron beam passes through the middle of the magnetic field in a direction perpendicular to the plane of the diagram (velocity

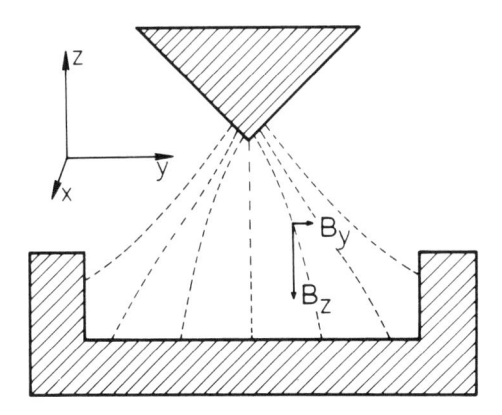

Fig. 1.2. Stern-Gerlach experiment with free electrons

$v = v_x$). The spins align parallel or antiparallel to the magnetic field and the electrons experience a deflecting force in the inhomogeneous field. In the plane of symmetry the force that tends to split the beam is

$$F = \pm\mu\frac{\partial B_z}{\partial z}, \tag{1.1}$$

where μ is the magnetic moment of the electrons. In addition, the electrons experience a Lorentz force due to their electric charge. Its component in the y direction, caused by the magnetic field component B_z, produces a right-hand shift of the image that could be detected by a photographic plate. As the electron beam has a certain width, it is also affected by the field component B_y which exists outside the symmetry plane. The com-

4   1. Introduction

ponent of the Lorentz force $F_L = (e/c)v_x B_y$, caused by B_y, deflects the electrons upwards if they are to the right of the symmetry plane and downwards if they are to the left of it. This causes a tilting of the traces on the photographic plate as is shown schematically in Fig. 1.3.

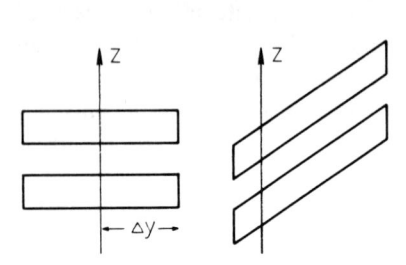

Fig. 1.3. Deflection of uncharged (left-hand side) and charged (right-hand side) particles with spin 1/2 in Stern-Gerlach field

Fig. 1.4. Transverse beam spread

Even in "thought" (Gedanken) experiments we must not consider an infinitely narrow beam, since the uncertainty principle must be taken into account, i.e., $\Delta y \cdot m \Delta v_y \approx h$. Because we want to work with a reasonable beam, the uncertainty of the velocity in the y direction Δv_y must be small compared to v_x (see Fig. 1.4). From this, together with the uncertainty relation given above, it follows that $h/m\Delta y \ll v_x$, or with $\lambda = h/mv_x$ (de Broglie wavelength)

$$\lambda \ll \Delta y ; \tag{1.2}$$

correspondingly one has $\lambda \ll \Delta z$.

Nevertheless, to be able to draw Fig. 1.5 clearly, we assume for now that we could have a beam whose spread in the z direction, in which we hope to obtain the splitting, is smaller than the de Broglie wavelength. Let us consider two points A' and B for which the y coordinate differs by λ. This is always possible since the beam width Δy is much greater than λ. As λ is small compared to the macroscopic dimensions of the field, the Taylor expansion

$$B_y(y + \lambda) = B_y(y) + \lambda \frac{\partial B_y}{\partial y}(y) \tag{1.3}$$

is, to a good approximation, valid. This means that the Lorentz force

1.2 Why Conventional Polarization Filters Do Not Work with Electrons

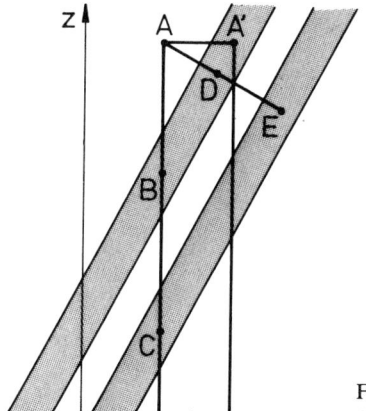

Fig. 1.5. Impossibility of the Stern-Gerlach experiment with free electrons

experienced by the electrons arriving at A' has always been larger by about $\Delta F_\text{L} = (e/c)v_x\lambda(\partial B_y/\partial y)$ than that experienced by the electrons arriving at B.

Thus A' is higher than B by an amount AB shown in Fig. 1.5. We can easily compare this distance with the splitting BC caused by the force F from Eq. (1.1). Since AB and BC are proportional to the respective forces applied, one obtains

$$\frac{\text{AB}}{\text{BC}} = \frac{\Delta F_\text{L}}{2F} = \frac{(e/c)v_x\lambda(\partial B_z/\partial z)}{2(eh/2mc)(\partial B_z/\partial z)} = \frac{2\pi\lambda}{\lambda} = 2\pi, \quad (1.4)$$

where use has been made of div $B = 0$, or $\partial B_y/\partial y = -\partial B_z/\partial z$. This means that the tilting of the traces is very large: AB is much larger than the splitting BC, although A'A is as small as λ. This has the following consequences:

If AE is the perpendicular from A to the traces, then, because AB > BC, AD is greater than DE. On the other hand, AD is smaller than the hypotenuse AA' $= \lambda$ of the right triangle ADA'; hence DE, the distance between the centers of the traces, is such that DE < AD < λ. This means that this distance is smaller than the width of either of the traces, which we have shown is considerably larger than λ in every direction. Consequently, no splitting into traces with opposing spin directions can be observed. The uncertainty principle, together with the Lorentz force, prevents spin-up and spin-down electrons from being separated by a macroscopic field of the Stern-Gerlach type. The most one could expect would be a slight imbalance of the spin directions at the edges of the beam.

Attempts have frequently been made to disprove the above argument, originating from BOHR and PAULI, that a Stern-Gerlach type experiment is impossible with electrons (see [1.2]). Such attempts have the same challenge as "thought" experiments for constructing perpetual-motion machines. However, all suggestions for modifying the experiment so that it would work have failed.

This does not mean that polarized free electrons cannot be produced at all. Scattering of unpolarized electrons by heavy atoms, for example, yields highly polarized electrons. In this way, however, one does not lose only a factor of 2, as with a conventional polarization filter, but a factor of 10^4 to 10^7, depending on how high a polarization one wants. As we shall see later, there are methods other than scattering, but they all have in common that they yield only small intensities. Nobody has yet found a spin filter for electrons that reduces the intensity by just a factor of 2.

For a polarization experiment one also needs an analyzer for the polarization. Here we have the same situation. If the transmission axis of an optical analyzer is parallel to the polarization, a totally polarized light beam passes through the analyzer without loss of intensity. Similarly, if one uses a spin filter of the Stern-Gerlach type as an analyzer, a totally polarized atomic beam passes through without appreciable loss of intensity, if the direction of its polarization is parallel to the analyzing direction. With electrons, however, one cannot use such a spin filter as an analyzer for the same reason one cannot use it as a polarizer. One must use some spin-dependent collision process, usually electron scattering, where one again loses several orders of magnitude in intensity.

Since one needs a polarizer as well as an analyzer for a polarization experiment, the two factors together easily make an intensity loss of a factor of 10^8 in an electron-polarization experiment. If we compare this to the factor of 2 for a light- or atom-polarization experiment (under ideal conditions), we see why electron-polarization studies became feasible only in recent years: Sufficiently advanced experimental techniques had to be developed before this field was accessible.

The fact that conventional polarization filters do not work with electrons does not mean that it is absolutely impossible to find effective electron polarization filters. As will be discussed in Subsection 6.1.2 there are interesting developments which show that it is worthwhile to search for "unconventional" electron polarization filters of high efficiency.

Before we can discuss quantitatively the processes in which electron polarization plays a role, we must look at the possibilities of describing polarized electrons mathematically.

2. Description of Polarized Electrons

2.1 A Few Results from Elementary Quantum Mechanics

The formal description of the spin of free electrons is summarized.

The following facts can be drawn from textbooks on quantum mechanics: The observable "spin" is represented by the operator s which satisfies the commutation relations characteristic of angular momenta:

$$s_x s_y - s_y s_x = i\hbar s_z \quad \text{(etc. by cyclic permutation)}. \tag{2.1}$$

If one separates out the factor $\hbar/2$ by the definition $s = (\hbar/2)\sigma$, one obtains from the above commutation relations, with the additional condition that σ_z is diagonal, the Pauli matrices

$$\sigma_x = \begin{pmatrix} 0 & 1 \\ 1 & 0 \end{pmatrix}, \quad \sigma_y = \begin{pmatrix} 0 & -i \\ i & 0 \end{pmatrix}, \quad \sigma_z = \begin{pmatrix} 1 & 0 \\ 0 & -1 \end{pmatrix}. \tag{2.2}$$

These operators receive their meaning from their application to the two-component wave functions $\begin{pmatrix} a_1 \\ a_2 \end{pmatrix}$ with whose help the two possible orientations of the electron spin can be described. For example, one has the eigenvalue equations

$$\sigma_z \begin{pmatrix} 1 \\ 0 \end{pmatrix} = \begin{pmatrix} 1 & 0 \\ 0 & -1 \end{pmatrix} \begin{pmatrix} 1 \\ 0 \end{pmatrix} = 1 \cdot \begin{pmatrix} 1 \\ 0 \end{pmatrix}, \quad \sigma_z \begin{pmatrix} 0 \\ 1 \end{pmatrix} = \begin{pmatrix} 1 & 0 \\ 0 & -1 \end{pmatrix} \begin{pmatrix} 0 \\ 1 \end{pmatrix} = -1 \cdot \begin{pmatrix} 0 \\ 1 \end{pmatrix} \tag{2.3}$$

which mean that $\begin{pmatrix} 1 \\ 0 \end{pmatrix}$ is an eigenfunction of σ_z with the eigenvalue $+1$ (or $+\hbar/2$ of s_z) and $\begin{pmatrix} 0 \\ 1 \end{pmatrix}$ belongs to the eigenvalue -1.

2. Description of Polarized Electrons

We can use these two states as a basis for representing the general state

$$\chi = \begin{pmatrix} a_1 \\ a_2 \end{pmatrix}$$

as a linear superposition

$$a_1 \begin{pmatrix} 1 \\ 0 \end{pmatrix} + a_2 \begin{pmatrix} 0 \\ 1 \end{pmatrix} = \begin{pmatrix} a_1 \\ a_2 \end{pmatrix}. \tag{2.4}$$

When χ is assumed to be normalized one has

$$\langle \chi | \chi \rangle = (a_1^*, a_2^*) \begin{pmatrix} a_1 \\ a_2 \end{pmatrix} = |a_1|^2 + |a_2|^2 = 1. \tag{2.5}$$

Remembering the quantum mechanical interpretation of the expansion of a wave function, we see from the left side of (2.4) that $|a_1|^2$ is the probability of finding the value $+\hbar/2$ when measuring the spin component in the z direction; that is, $|a_1|^2$ is the probability of finding the electron in the state $\begin{pmatrix} 1 \\ 0 \end{pmatrix}$. $|a_2|^2$ is the probability of finding the eigenvalue $-\hbar/2$, that is, of finding the state $\begin{pmatrix} 0 \\ 1 \end{pmatrix}$. A measurement of the spin direction forces the electron with the probability $|a_i|^2$ ($i = 1, 2$) into one of the two eigenstates. Even if the spins have been oriented in the x direction, by measuring the spin components in the z direction the values $+\hbar/2$ or $-\hbar/2$ are obtained, each with the probability $1/2$ (see Sect. 2.2). Thus the spins are seen to be affected by the measurement.

This disturbance of the spin state by measurement makes it impossible to measure all the spin components simultaneously. This follows mathematically from the noncommutativity of the angular momentum components, (2.1). The operator s^2, however, which has the eigenvalue $s(s+1)\hbar^2 = 3/4\hbar^2$, commutes with the components s_x, s_y and s_z. One can measure its eigenvalue simultaneously with those of any of the components of s. For these reasons, the statement that "the spin is in the z direction", means more precisely the following: The spin vector lies somewhere on a conical shell in such a way that its component in the z direction is $\hbar/2$; the two other components are not known, it is merely known that $s_x^2 + s_y^2 + s_z^2 = 3/4\hbar^2$ (see Fig. 2.1).

It follows from the above statements that one cannot distinguish between electron beams in which all directions are equally likely and those in which half of the spins are parallel and half are antiparallel to some

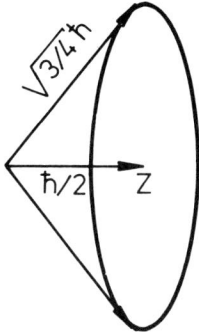

Fig. 2.1. Spin "in the z direction"

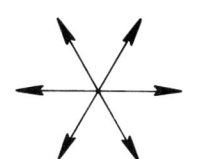

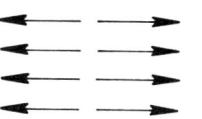

Fig. 2.2. Unpolarized electrons

arbitrary reference direction (see Fig. 2.2). This is because in both cases a measurement results in half of the spins being parallel and half being antiparallel to the direction specified by the observation. Every conceivable experiment with such electron beams yields the same result; the beams therefore must be regarded as identical.

2.2 Pure Spin States

The spin function describing the spin in an arbitrary direction is determined. The polarization is defined as the expectation value of the spin operator. Its magnitude for a pure state is 1.

In this section we consider electrons which are all in the same spin state. In such cases the system of electrons is said to be in a pure spin state.

The spin direction of a state which is described by $\chi = \begin{pmatrix} a_1 \\ a_2 \end{pmatrix}$ is specified by a_1 and a_2, as will now be shown. Let $\hat{e} = (e_x, e_y, e_z)$ be the unit vector in the direction ϑ, φ, i.e. (see Fig. 2.3),

$$e_x = \sin\vartheta \cos\varphi, \quad e_y = \sin\vartheta \sin\varphi, \quad e_z = \cos\vartheta.$$

We now ask what the spin function that describes a spin in the direction ϑ, φ would look like. For this we must solve the eigenvalue equation $(\boldsymbol{\sigma} \cdot \hat{e})\chi = \lambda\chi$, since $\boldsymbol{\sigma} \cdot \hat{e}$ is the projection of the spin operator in the specified direction. Since

$$\sigma_x \begin{pmatrix} a_1 \\ a_2 \end{pmatrix} = \begin{pmatrix} a_2 \\ a_1 \end{pmatrix}, \quad \sigma_y \begin{pmatrix} a_1 \\ a_2 \end{pmatrix} = \begin{pmatrix} -ia_2 \\ ia_1 \end{pmatrix}, \quad \sigma_z \begin{pmatrix} a_1 \\ a_2 \end{pmatrix} = \begin{pmatrix} a_1 \\ -a_2 \end{pmatrix} \quad (2.6)$$

2. Description of Polarized Electrons

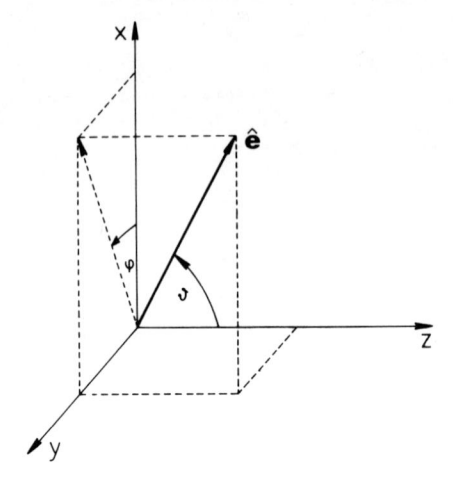

Fig. 2.3. Spin direction

one obtains

$$(\boldsymbol{\sigma} \cdot \hat{e})\chi = \begin{pmatrix} a_2 \sin \vartheta \cos \varphi - i a_2 \sin \vartheta \sin \varphi + a_1 \cos \vartheta \\ a_1 \sin \vartheta \cos \varphi + i a_1 \sin \vartheta \sin \varphi - a_2 \cos \vartheta \end{pmatrix}$$
$$= \begin{pmatrix} a_1 \cos \vartheta + a_2 \sin \vartheta e^{-i\varphi} \\ a_1 \sin \vartheta e^{i\varphi} - a_2 \cos \vartheta \end{pmatrix}.$$

The attempt to find the eigenfunction of the component of the spin operator in the $\hat{e}$ direction, i.e., to solve the equation $(\boldsymbol{\sigma} \cdot \hat{e})\chi = \lambda \chi$, thus gives

$$a_1(\cos \vartheta - \lambda) + a_2 \sin \vartheta e^{-i\varphi} = 0$$
$$a_1 \sin \vartheta e^{i\varphi} + a_2(-\cos \vartheta - \lambda) = 0. \tag{2.7}$$

The condition for a non-trivial solution (disappearance of the determinant) is

$$-\cos^2 \vartheta + \lambda^2 - \sin^2 \vartheta = 0 \quad \text{or} \quad \lambda = \pm 1.$$

For $\lambda = +1$, one obtains from (2.7)

$$\frac{a_2}{a_1} = \frac{\cos \vartheta - 1}{-\sin \vartheta e^{-i\varphi}} = \tan \frac{\vartheta}{2} e^{i\varphi}, \tag{2.8}$$

and for $\lambda = -1$

$$\frac{a_2}{a_1} = \frac{\cos \vartheta + 1}{-\sin \vartheta e^{-i\varphi}} = -\cot \frac{\vartheta}{2} e^{i\varphi}. \tag{2.9}$$

2.2 Pure Spin States

Since a_1 and a_2 are solutions of a homogeneous system of equations, they are determined except for a constant. This constant can be specified by normalization according to (2.5). Then

$$a_1 = \cos\frac{\vartheta}{2}, \quad a_2 = \sin\frac{\vartheta}{2} e^{i\varphi} \quad \text{for} \quad \lambda = +1 \tag{2.10}$$

$$a_1 = \sin\frac{\vartheta}{2}, \quad a_2 = -\cos\frac{\vartheta}{2} e^{i\varphi} \quad \text{for} \quad \lambda = -1. \tag{2.11}$$

A common phase factor of a_1 and a_2 which remains undetermined has been chosen arbitrarily.

The spin functions with the a_1 and a_2 just calculated are eigenfunctions of the component of the spin operator σ in the $\hat{e}$ direction with the eigenvalues $+1$ and -1, that is, they represent the states where the spin in the direction ϑ, φ has the value $+\hbar/2$ or $-\hbar/2$. One can immediately see that the solutions (2.9) or (2.11) are none other than the solutions (2.8) or (2.10), respectively, for the direction $-\hat{e}$ which is described by the angles $\pi - \vartheta$, $\varphi + \pi$. Thus it is sufficient from now on to use only the solutions (2.8) or (2.10).

In the special cases $\vartheta = 0$ or π, $\varphi = 0$ (spin parallel or antiparallel to the z direction), Eq. (2.10) yields the expected eigenfunctions of σ_z, $\begin{pmatrix} 1 \\ 0 \end{pmatrix}$ or $\begin{pmatrix} 0 \\ 1 \end{pmatrix}$. For $\vartheta = \pi/2$, $\varphi = 0$ (spin in the x direction) one obtains the spin function $\chi = \begin{pmatrix} 1/\sqrt{2} \\ 1/\sqrt{2} \end{pmatrix}$. The latter example should forewarn us of a false conclusion: A superposition of spin states in opposing directions with equal amplitudes and with a fixed phase relation, such as

$$\frac{1}{\sqrt{2}}\begin{pmatrix} 1 \\ 0 \end{pmatrix} + \frac{1}{\sqrt{2}}\begin{pmatrix} 0 \\ 1 \end{pmatrix} = \begin{pmatrix} 1/\sqrt{2} \\ 1/\sqrt{2} \end{pmatrix}$$

does not produce a cancellation of spins, but a spin in another direction. This is analogous to a coherent superposition of right- and left-circularly polarized light waves with a fixed phase relation, which also does not produce an unpolarized wave but a linearly polarized wave.

Let us now consider the polarization of the electrons described by $\begin{pmatrix} a_1 \\ a_2 \end{pmatrix}$. Whereas the eigenvalue represents the result of a single measurement, the polarization tells us something about the average spin direction of the ensemble. It is therefore an expectation value.

2. Description of Polarized Electrons

We recall that the expectation value is the average over all values which the considered property of a particle can assume in a given state ψ. A single measurement of a property, described by the operator Q, yields an eigenvalue. If such measurements are made on a large number of identical systems one generally observes all possible eigenvalues. The average of the eigenvalues thus obtained is the expectation value

$$\langle Q \rangle = \langle \psi | Q | \psi \rangle. \tag{2.12}$$

For example, for the hydrogen atom in the ground state ψ_0, the expectation value of the momentum $\boldsymbol{p}$ is clearly zero as the momentum of the orbiting electron constantly changes its direction. In fact, one has

$$\langle \boldsymbol{p} \rangle = \langle \psi_0 | \boldsymbol{p} | \psi_0 \rangle = -i\hbar \int \psi_0^* \, \mathrm{grad}\, \psi_0 \, d\tau = 0,$$

as can easily be seen by using the simple wave function of the ground state.

If a system is in the eigenstate of an operator each measurement of the corresponding observable will definitely yield the eigenvalue, so that the expectation value coincides with the eigenvalue. For example, for a system in a normalized energy state ψ_n with the eigenvalue E_n, one obtains the expectation value

$$\langle H \rangle = \langle \psi_n | H | \psi_n \rangle = \langle \psi_n | E_n | \psi_n \rangle = E_n \langle \psi_n | \psi_n \rangle = E_n.$$

After this brief look at elementary quantum mechanics, we return to the polarization and define it as the expectation value of the Pauli spin operator

$$\boldsymbol{P} = \langle \boldsymbol{\sigma} \rangle = \langle \chi | \boldsymbol{\sigma} | \chi \rangle = (a_1^*, a_2^*) \boldsymbol{\sigma} \begin{pmatrix} a_1 \\ a_2 \end{pmatrix}. \tag{2.13}$$

With this definition one finds from (2.6) and (2.10) that the components of the polarization vector are

$$\left. \begin{aligned} P_x &= (a_1^*, a_2^*) \begin{pmatrix} a_2 \\ a_1 \end{pmatrix} \\ &= \cos\frac{\vartheta}{2} \sin\frac{\vartheta}{2} e^{i\varphi} + \cos\frac{\vartheta}{2} \sin\frac{\vartheta}{2} e^{-i\varphi} = \sin\vartheta \cos\varphi \\ P_y &= (a_1^*, a_2^*) \begin{pmatrix} -ia_2 \\ ia_1 \end{pmatrix} \\ &= \cos\frac{\vartheta}{2} \sin\frac{\vartheta}{2} \frac{1}{i} e^{i\varphi} - \cos\frac{\vartheta}{2} \sin\frac{\vartheta}{2} \frac{1}{i} e^{-i\varphi} = \sin\vartheta \sin\varphi \\ P_z &= (a_1^*, a_2^*) \begin{pmatrix} a_1 \\ -a_2 \end{pmatrix} = \cos^2\frac{\vartheta}{2} - \sin^2\frac{\vartheta}{2} = \cos\vartheta. \end{aligned} \right\} \tag{2.14}$$

It can be seen from these equations that the polarization has the direction

ϑ, φ and that the degree of polarization which is defined by

$$P = \sqrt{P_x^2 + P_y^2 + P_z^2}$$

is 1 in the case discussed here. This is reasonable, as we have assumed that the electron spins can be described by a single spin function $\begin{pmatrix} a_1 \\ a_2 \end{pmatrix}$ (pure state) so that there is only one spin direction in the beam, namely that in the direction ϑ, φ specified in (2.10).

If the state $\chi = \begin{pmatrix} a_1 \\ a_2 \end{pmatrix}$ is not normalized, a sensible extension of the definition (2.13) is

$$\boldsymbol{P} = \frac{\langle \chi | \boldsymbol{\sigma} | \chi \rangle}{\langle \chi | \chi \rangle}. \tag{2.15}$$

Thus the magnitude of the components of $\boldsymbol{P}$ remains between 0 and 1, e.g.,

$$P_z = \frac{|a_1|^2 - |a_2|^2}{|a_1|^2 + |a_2|^2}.$$

The polarization vector can be completely determined. One can measure all its components, but one must take care not to use the same particles for this, since the states of the particles are affected by a measurement, so that a subsequent measurement could give a wrong value. One might, for example, with a beam that is polarized in the x direction [polarization $\boldsymbol{P} = (1, 0, 0)$], proceed as follows. First, one could measure the spin components in the z direction. Half of the measurements would yield $+\hbar/2$, the other half $-\hbar/2$, which would imply $P_z = 0$. If one then carried out a measurement in the y or x direction on the same electrons which would then have an equal number of spins parallel and antiparallel to the z direction, one would again obtain $+\hbar/2$ and $-\hbar/2$ with equal probability, implying $P_x = 0$, $P_y = 0$. If the measurement had been carried out properly, however, in the x direction one should have found only the value $+\hbar/2$, since we initially assumed the beam to be totally polarized in this direction.

The objection that one should have started with the measurement in the x direction is not valid, as one does not know before making the measurement that the beam is polarized in this direction (otherwise the measurement would be superfluous).

In order to conduct the experiment properly, one must make the measurements on different subsystems of the electron ensemble. In doing so, one must of course be sure that the subsystems are in the same polarization state as the total system. For example, one can make the measurements sequentially on a beam of constant polarization. One then always uses electrons which have not been affected by a previous measurement and obtains, in the above example, $P_x = 1$, $P_y = 0$, and $P_z = 0$.

2.3 Statistical Mixtures of Spin States. Description of Electron Polarization by Density Matrices

Partially polarized beams represent a statistical mixture of different spin states. To describe them, one can suitably apply density matrices. The connection between polarization and density matrices is given.

Until now, only totally polarized electron beams have been considered, that is, ensembles in which all particles are in the same spin state. Now we will consider partially polarized beams, which are statistical mixtures of spin states. In this case, the polarization of the total system is the average of the polarization values $\boldsymbol{P}^{(n)}$ of the individual systems which are in pure spin states $\chi^{(n)}$:

$$\boldsymbol{P} = \sum_n g^{(n)} \boldsymbol{P}^{(n)} = \sum_n g^{(n)} \langle \chi^{(n)} | \boldsymbol{\sigma} | \chi^{(n)} \rangle, \qquad (2.16)$$

where the weighting factors $g^{(n)}$ take into account the relative proportion of the states $\chi^{(n)}$:

$$g^{(n)} = \frac{N^{(n)}}{\sum_n N^{(n)}},$$

where $N^{(n)}$ is the number of electrons in the state $\chi^{(n)}$. The $\chi^{(n)}$ have been assumed to be normalized here.

As an expedient means of describing the polarization in this case, one can use the density matrix ρ which is defined as [2.1–3]

$$\rho = \sum_n g^{(n)} \begin{pmatrix} |a_1^{(n)}|^2 & a_1^{(n)} a_2^{(n)*} \\ a_1^{(n)*} a_2^{(n)} & |a_2^{(n)}|^2 \end{pmatrix}. \qquad (2.17)$$

The individual matrices of this sum are the density matrices of the pure states.

The density matrix is connected to the polarization by the relation

$$\boldsymbol{P} = \operatorname{tr} \rho \boldsymbol{\sigma} \quad \text{or} \quad P_i = \operatorname{tr} \rho \sigma_i. \qquad (2.18)$$

2.3 Statistical Mixtures of Spin States

By using (2.2) one sees immediately that these equations are correct. For example, we have

$$\rho\sigma_x = \sum_n g^{(n)} \begin{pmatrix} a_1^{(n)} a_2^{(n)*} & |a_1^{(n)}|^2 \\ |a_2^{(n)}|^2 & a_1^{(n)*} a_2^{(n)} \end{pmatrix},$$

and thus

$$\operatorname{tr} \rho\sigma_x = \sum_n g^{(n)} \{a_1^{(n)} a_2^{(n)*} + a_1^{(n)*} a_2^{(n)}\} = \sum_n g^{(n)} P_x^{(n)} = P_x,$$

where use has been made of part of (2.14). Similarly

$$\operatorname{tr} \rho\sigma_z = \sum_n g^{(n)} \{|a_1^{(n)}|^2 - |a_2^{(n)}|^2\} = \sum_n g^{(n)} P_z^{(n)} = P_z.$$

One can express the elements of the density matrix in terms of the components of the polarization. Using (2.14), (2.17), and (2.2) one then obtains

$$\rho = \frac{1}{2}\begin{pmatrix} 1 + P_z & P_x - iP_y \\ P_x + iP_y & 1 - P_z \end{pmatrix} = \frac{1}{2}(\mathbf{1} + \mathbf{P}\cdot\boldsymbol{\sigma}), \tag{2.19}$$

where **1** is the unit matrix.

In making the definition (2.16) we assumed that the states $\chi^{(n)}$ were normalized; the relative proportions of the single states were taken into account by using weighting factors. One can also start with unnormalized $\chi^{(n)}$. Then weighting factors are unnecessary, since the relative proportion of the nth state is already expressed by the unnormalized amplitude of $\chi^{(n)}$; it is given by

$$\frac{\langle \chi^{(n)} | \chi^{(n)} \rangle}{\sum_n \langle \chi^{(n)} | \chi^{(n)} \rangle}.$$

In this case the polarization is

$$\mathbf{P} = \frac{\sum_n \langle \chi^{(n)} | \boldsymbol{\sigma} | \chi^{(n)} \rangle}{\sum_n \langle \chi^{(n)} | \chi^{(n)} \rangle} \tag{2.20}$$

(which can also be written as

$$\mathbf{P} = \frac{\sum_n \langle \chi^{(n)} | \chi^{(n)} \rangle [\langle \chi^{(n)} | \boldsymbol{\sigma} | \chi^{(n)} \rangle / \langle \chi^{(n)} | \chi^{(n)} \rangle]}{\sum_n \langle \chi^{(n)} | \chi^{(n)} \rangle}$$

thus leading back to the form (2.16) with normalized functions). Instead of (2.18) we then have

$$\mathbf{P} = \frac{\operatorname{tr} \rho \boldsymbol{\sigma}}{\operatorname{tr} \rho}, \qquad (2.21)$$

where the density matrix has the form

$$\rho = \sum_n \begin{pmatrix} |a_1^{(n)}|^2 & a_1^{(n)} a_2^{(n)*} \\ a_1^{(n)*} a_2^{(n)} & |a_2^{(n)}|^2 \end{pmatrix}. \qquad (2.22)$$

The denominator

$$\sum_n \langle \chi^{(n)} | \chi^{(n)} \rangle = \sum_n \{|a_1^{(n)}|^2 + |a_2^{(n)}|^2\} = \operatorname{tr} \rho$$

now appearing in the polarization formulae (2.20) and (2.21) must also be taken into account on the left-hand side of (2.19), so that the corresponding relation is

$$\frac{\rho}{\operatorname{tr} \rho} = \frac{1}{2} \begin{pmatrix} 1 + P_z & P_x - iP_y \\ P_x + iP_y & 1 - P_z \end{pmatrix} = \frac{1}{2}(1 + \mathbf{P} \cdot \boldsymbol{\sigma}). \qquad (2.23)$$

The density matrix assumes its simplest form if one takes the direction of the resultant polarization as the z axis of the coordinate system shown in Fig. 2.3, i.e., chooses $P_x = P_y = 0$, $P = P_z$. Then from (2.19) (if we return to normalized states) one has

$$\rho = \frac{1}{2} \begin{pmatrix} 1 + P & 0 \\ 0 & 1 - P \end{pmatrix}. \qquad (2.24)$$

The density matrix thus is transformed to a diagonal form.

This form of the density matrix illustrates again the meaning of P: Since $|a_1^{(n)}|^2$ is the probability that the eigenvalue $+\hbar/2$ will be obtained from a spin measurement in the z direction on the nth subsystem, the probability is $\sum_n g^{(n)} |a_1^{(n)}|^2$ that this measurement on the total beam will give the value $+\hbar/2$. This probability can also be expressed as $N_\uparrow/(N_\uparrow + N_\downarrow)$, where $N_\uparrow$ is the number of measurements that yield the value $+\hbar/2$ and $N_\uparrow + N_\downarrow$ is the total number of measurements. (Correspondingly, $\sum_n g^{(n)} |a_2^{(n)}|^2 = N_\downarrow/(N_\uparrow + N_\downarrow)$ is the probability that the value $-\hbar/2$ will be obtained). Thus one has $N_\uparrow/(N_\uparrow + N_\downarrow) = \sum_n g^{(n)} |a_1^{(n)}|^2 = \frac{1}{2}(1 + P)$, where the last part of the equation comes from a comparison of (2.17) and

2.3 Statistical Mixtures of Spin States

(2.24). Consequently, one obtains for the degree of polarization

$$P = \frac{N_\uparrow - N_\downarrow}{N_\uparrow + N_\downarrow}. \tag{2.25}$$

For a beam totally polarized in the $+z$ direction ($N_\downarrow = 0$), the diagonal form of the density matrix (2.24) becomes

$$\rho = \begin{pmatrix} 1 & 0 \\ 0 & 0 \end{pmatrix}. \tag{2.26}$$

For an unpolarized beam ($N_\uparrow = N_\downarrow$) one obtains

$$\rho = \begin{pmatrix} \tfrac{1}{2} & 0 \\ 0 & \tfrac{1}{2} \end{pmatrix}. \tag{2.27}$$

From the identity

$$\rho = \frac{1}{2}\begin{pmatrix} 1+P & 0 \\ 0 & 1-P \end{pmatrix} = (1-P)\begin{pmatrix} \tfrac{1}{2} & 0 \\ 0 & \tfrac{1}{2} \end{pmatrix} + P\begin{pmatrix} 1 & 0 \\ 0 & 0 \end{pmatrix} \tag{2.28}$$

one sees by comparison with (2.26) and (2.27) that an electron beam with an arbitrary degree of polarization P can be considered to be made up of a totally polarized fraction and an unpolarized fraction which are mixed in the ratio $P/(1-P)$.

We will illustrate the general definitions introduced here by two simple examples.

Example 2.1: In an ensemble of 100 electrons, one finds 80 electrons with spin $+\hbar/2$ in the z direction and 20 with $-\hbar/2$. With $N_\uparrow = 80$, $N_\downarrow = 20$, one has, from (2.25),

$$P = 0.6.$$

From (2.28), this can also be expressed by saying that 60% of the beam is totally polarized and 40% is unpolarized (see Fig. 2.4).

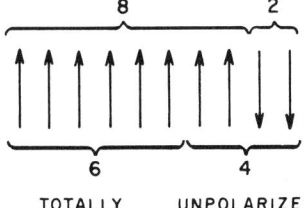

Fig. 2.4. Partially polarized beam

Example 2.2: An electron beam, which is totally polarized in the z direction, is mixed with another beam of the same intensity, which is totally polarized in the x direction. According to (2.16) the resulting polarization is then (see Fig. 2.5)

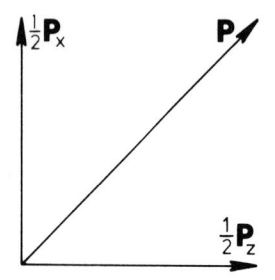

$$P = \tfrac{1}{2}P_x + \tfrac{1}{2}P_z,$$

Fig. 2.5. Superposition of polarization vectors

forming a 45° angle with both the z and the x direction. Its magnitude is $1/\sqrt{2} = 0.7071$. One has $P < 1$ because the two component beams are independent of each other so that they are incoherently superimposed. (With opposing spin directions for the two beams one would have $P = 0$).

This result can also be obtained by using the density matrix: From (2.10) the two component beams have the normalized eigenfunctions $\begin{pmatrix} 1 \\ 0 \end{pmatrix}$ and $\begin{pmatrix} \cos 45° \\ \sin 45° \end{pmatrix} = \begin{pmatrix} 1/\sqrt{2} \\ 1/\sqrt{2} \end{pmatrix}$ so that from (2.17) the density matrix is

$$\rho = \frac{1}{2}\begin{pmatrix} 1 & 0 \\ 0 & 0 \end{pmatrix} + \frac{1}{2}\begin{pmatrix} \tfrac{1}{2} & \tfrac{1}{2} \\ \tfrac{1}{2} & \tfrac{1}{2} \end{pmatrix} = \frac{1}{2}\begin{pmatrix} \tfrac{3}{2} & \tfrac{1}{2} \\ \tfrac{1}{2} & \tfrac{1}{2} \end{pmatrix}.$$

From this and (2.19) it follows that

$$P_z = \frac{3}{2} - 1 = \frac{1}{2}, \quad P_y = 0, \quad P_x = \frac{1}{2}$$

and

$$P = \sqrt{P_x^2 + P_z^2} = 1/\sqrt{2}.$$

By proper choice of the coordinate system the density matrix could have been transformed to the diagonal form (see Problem 2.1 below).

2.3 Statistical Mixtures of Spin States

In such simple cases one can, of course, obtain the result more quickly, without density matrices. Their real use becomes obvious in more complicated cases (see Sects. 3.3 and 5.2).

We emphasize once more the difference between coherent and incoherent superposition of spin states. The superposition of squares of amplitudes when forming the density matrix of a mixed state implies a loss of phase relations. The simplest example of such an incoherent superposition is the superposition of two opposing spin states with equal weighting factors, which yields $P = 0$. In contrast to this, with the coherent superposition of amplitudes, as mentioned in Section 2.2, the phase relations between the amplitudes are retained and one obtains a completely polarized state. This is because every new state

$$\begin{pmatrix} \sum_n a_1^{(n)} \\ \sum_n a_2^{(n)} \end{pmatrix} = \begin{pmatrix} A_1 \\ A_2 \end{pmatrix}$$

formed in this way is an eigenfunction of the component $(\sigma \cdot \hat{e})$ of the spin operator in the direction ϑ, φ which is specified by $A_2/A_1 = \tan\frac{\vartheta}{2} \exp(i\varphi)$ (cf. Sect. 2.2). (Every complex number $A_2/A_1 = |R| \exp(i\varphi)$ can be represented in this way as the tangent passes through all real numbers). Thus the polarization in this direction is

$$P = \frac{(A_1^*, A_2^*)(\sigma \cdot \hat{e})\begin{pmatrix} A_1 \\ A_2 \end{pmatrix}}{(A_1^*, A_2^*)\begin{pmatrix} A_1 \\ A_2 \end{pmatrix}} = 1, \quad \text{since} \quad (\sigma \cdot \hat{e})\begin{pmatrix} A_1 \\ A_2 \end{pmatrix} = 1 \cdot \begin{pmatrix} A_1 \\ A_2 \end{pmatrix}.$$

Problem 2.1: For Example 2.2, give the density matrix in diagonal form using the degree of polarization calculated there. Check the result by using the spin functions.

Solution: As the magnitude of the polarization was found above to be $P = 0.7071$ the density matrix in this case must be, according to (2.24),

$$\rho = \frac{1}{2}\begin{pmatrix} 1.7071 & 0 \\ 0 & 0.2929 \end{pmatrix}.$$

We will check to see if this is correct: In the coordinate system, where the z axis is in the direction of the resultant polarization vector, the spin directions of the two constituent beams have the angles $\vartheta = 45°$, and $\varphi = 0°$ and $180°$, respectively. Their spin functions, from (2.10), are

$$\begin{pmatrix} \cos\frac{\vartheta}{2} \\ \sin\frac{\vartheta}{2} \end{pmatrix} \text{ and } \begin{pmatrix} \cos\frac{\vartheta}{2} \\ \sin\frac{\vartheta}{2} e^{i\pi} \end{pmatrix} \quad \text{with} \quad \vartheta = 45°.$$

2. Description of Polarized Electrons

Thus the density matrix of the total system has the expected diagonal form

$$\rho = \frac{1}{2}\rho^{(1)} + \frac{1}{2}\rho^{(2)} = \frac{1}{2}\begin{pmatrix} 2\cos^2\frac{\vartheta}{2} & \cos\frac{\vartheta}{2}\sin\frac{\vartheta}{2}(1+e^{-i\pi}) \\ \cos\frac{\vartheta}{2}\sin\frac{\vartheta}{2}(1+e^{i\pi}) & 2\sin^2\frac{\vartheta}{2} \end{pmatrix}$$

$$= \frac{1}{2}\begin{pmatrix} 1.7071 & 0 \\ 0 & 0.2929 \end{pmatrix} \quad \text{for} \quad \vartheta = 45°.$$

3. Polarization Effects in Electron Scattering from Unpolarized Targets

3.1 The Dirac Equation and Its Interpretation

By linearizing the relativistic generalization of the Schrödinger equation, one obtains the Dirac equation. It is Lorentz invariant and describes the electron spin and spin-orbit coupling without the need to introduce further assumptions. The definition of the polarization as the expectation value of the spin operator is not Lorentz invariant and will therefore be referred to the rest system of the electrons.

In analogy to the case of light beams, electron beams can be polarized by scattering, and the angular distribution of scattered electrons depends on the state of polarization of the incident beam. These effects can be treated by the Dirac equation, which is the basic equation for describing the electron, including its spin and its relativistic behavior.

Dirac discovered this equation in 1928 when he tried to find a relativistic generalization of the Schrödinger equation. We can best see how the relativistic generalization can be made by recalling the path which formally leads to the Schrödinger equation.

One starts from the Hamiltonian function for a free particle

$$H = \frac{p^2}{2m} \tag{3.1}$$

and substitutes for p and H the operators

$$p = -i\hbar \nabla, \quad H = i\hbar \frac{\partial}{\partial t}. \tag{3.2}$$

By applying these to a wave function $\psi(r, t)$, one obtains the Schrödinger equation

$$i\hbar \dot{\psi} = -\frac{\hbar^2}{2m} \Delta \psi. \tag{3.3}$$

Does this method also lead to a useful result in the relativistic case? To examine this question we start with the relativistic energy law

$$H^2 = c^2 p^2 + m^2 c^4 \tag{3.4}$$

(m = rest mass). By substituting the operators (3.2) for p and H, we obtain

$$\left(\Delta - \frac{1}{c^2} \frac{\partial^2}{\partial t^2} - \frac{1}{\lambdabar^2} \right) \psi = 0, \tag{3.5}$$

where

$$\lambdabar = \frac{\hbar}{mc} \tag{3.6}$$

is the Compton wavelength.

When we include electric and magnetic potentials ϕ and A in the consideration, the Hamiltonian function for the non-relativistic case is

$$H = \frac{[p - (\varepsilon/c)A]^2}{2m} + \varepsilon\phi, \tag{3.7}$$

and thus

$$H - \varepsilon\phi = \frac{[p - (\varepsilon/c)A]^2}{2m},$$

where ε = electron charge = $-e$. This follows from (3.1), if one substitutes $p - (\varepsilon/c)A$ for p (p = canonical momentum) and $H - \varepsilon\phi$ for H. Correspondingly, it follows from (3.4) for the relativistic case

$$(H - \varepsilon\phi)^2 = (cp - \varepsilon A)^2 + m^2 c^4. \tag{3.8}$$

By interpreting H and p as operators, one obtains a wave equation for an electron in an external electromagnetic field (Klein-Gordon equation).

Serious difficulties are encountered in the use of this equation. For example, it predicts far too large a fine-structure splitting of the hydrogen spectrum. It is also problematic in its mathematical structure as a second-order differential equation in t, since one requires the initial values of ψ and $\dot\psi$ to solve it. The Schrödinger equation requires only the initial value of ψ, and it is difficult to see why the consideration of relativistic effects should lead to such radical differences in the initial information required to describe the behavior of the electron.

Dirac had the idea of splitting up the equation into a product of two linear expressions and of considering these individually. The equation

$$(H^2 - c^2 \sum_\mu p_\mu^2 - m^2 c^4)\psi = 0 \tag{3.9}$$

($p_\mu = p_x, p_y, p_z$, components of the momentum operator), which follows from the force-free form of (3.8), can be expressed in the form

$$(H - c \sum_\mu \alpha_\mu p_\mu - \beta m c^2)(H + c \sum_\mu \alpha_\mu p_\mu + \beta m c^2)\psi = 0 \tag{3.10}$$

if the constant coefficients α_μ and β satisfy the relations

$$\alpha_\mu \alpha_{\mu'} + \alpha_{\mu'} \alpha_\mu = 2\delta_{\mu\mu'}$$
$$\alpha_\mu \beta + \beta \alpha_\mu = 0 \tag{3.11}$$
$$\beta^2 = 1 \ .$$

This can easily be seen by multiplication. If one can solve the equation

$$(H - c \sum_\mu \alpha_\mu p_\mu - \beta m c^2)\psi = 0 \tag{3.12}$$

(or the corresponding equation from (3.10) with the plus sign) then (3.10) is also solved. The linearized equation (3.12) has the advantage that it is of the first order in $\partial/\partial t$ just as the Schrödinger equation is. The derivatives with respect to the space coordinates are of the same order, which is necessary for relativistic covariance.

The relations (3.11) cannot be satisfied with ordinary numbers. One can, however, solve (3.11) with matrices (at least 4×4), for example with

$$\alpha_x = \begin{pmatrix} 0 & 0 & 0 & 1 \\ 0 & 0 & 1 & 0 \\ 0 & 1 & 0 & 0 \\ 1 & 0 & 0 & 0 \end{pmatrix}, \quad \alpha_y = \begin{pmatrix} 0 & 0 & 0 & -i \\ 0 & 0 & i & 0 \\ 0 & -i & 0 & 0 \\ i & 0 & 0 & 0 \end{pmatrix},$$
$$\alpha_z = \begin{pmatrix} 0 & 0 & 1 & 0 \\ 0 & 0 & 0 & -1 \\ 1 & 0 & 0 & 0 \\ 0 & -1 & 0 & 0 \end{pmatrix}, \quad \beta = \begin{pmatrix} 1 & 0 & 0 & 0 \\ 0 & 1 & 0 & 0 \\ 0 & 0 & -1 & 0 \\ 0 & 0 & 0 & -1 \end{pmatrix}. \tag{3.13}$$

Thus the Dirac equation for a free particle is, if one arbitrarily chooses the left factor of (3.10),[1]

$$\left\{i\hbar\frac{\partial}{\partial t} + i\hbar c\left(\alpha_x\frac{\partial}{\partial x} + \alpha_y\frac{\partial}{\partial y} + \alpha_z\frac{\partial}{\partial z}\right) - \beta mc^2\right\}\psi = 0. \tag{3.14}$$

As α_μ and β are 4×4 matrices, the formula makes sense only when ψ has the form

$$\psi = \begin{pmatrix} \psi_1 \\ \psi_2 \\ \psi_3 \\ \psi_4 \end{pmatrix}.$$

Then (3.14) represents a system of four simultaneous first-order partial differential equations:

$$\begin{aligned}
i\hbar\frac{\partial}{\partial t}\psi_1 + i\hbar c\left(\frac{\partial}{\partial x}\psi_4 - i\frac{\partial}{\partial y}\psi_4 + \frac{\partial}{\partial z}\psi_3\right) - mc^2\psi_1 &= 0 \\
i\hbar\frac{\partial}{\partial t}\psi_2 + i\hbar c\left(\frac{\partial}{\partial x}\psi_3 + i\frac{\partial}{\partial y}\psi_3 - \frac{\partial}{\partial z}\psi_4\right) - mc^2\psi_2 &= 0 \\
i\hbar\frac{\partial}{\partial t}\psi_3 + i\hbar c\left(\frac{\partial}{\partial x}\psi_2 - i\frac{\partial}{\partial y}\psi_2 + \frac{\partial}{\partial z}\psi_1\right) + mc^2\psi_3 &= 0 \\
i\hbar\frac{\partial}{\partial t}\psi_4 + i\hbar c\left(\frac{\partial}{\partial x}\psi_1 + i\frac{\partial}{\partial y}\psi_1 - \frac{\partial}{\partial z}\psi_2\right) + mc^2\psi_4 &= 0.
\end{aligned} \tag{3.15}$$

Let us now turn to the description of free electrons by the Dirac equation. We set the z axis in the direction of propagation and take the wave function to be

$$\psi_j = a_j\, e^{i(kz-\omega t)}, \tag{3.16}$$

where the a_j must be determined such that equations (3.15) are solved. Substituting (3.16) into (3.15) and using the abbreviations $E = \hbar\omega$ and $p_z = \hbar k$, one obtains

$$\begin{aligned}
(E - mc^2)a_1 \qquad\qquad\qquad -cp_z a_3 \qquad\qquad\qquad &= 0 \\
(E - mc^2)a_2 \qquad\qquad\qquad +cp_z a_4 &= 0 \\
-cp_z a_1 \qquad\qquad +(E + mc^2)a_3 \qquad\qquad\qquad &= 0 \\
cp_z a_2 \qquad\qquad\qquad +(E + mc^2)a_4 &= 0.
\end{aligned} \tag{3.17}$$

---

[1] If one starts from the right factor as was usual in earlier literature, the two pairs of components ψ_1, ψ_2 and ψ_3, ψ_4 are interchanged. This choice is made in MOTT and MASSEY [3.1].

As a homogeneous linear system of equations, (3.17) has only a non-trivial solution if the determinant

$$\begin{vmatrix} (E - mc^2) & 0 & -cp_z & 0 \\ 0 & (E - mc^2) & 0 & cp_z \\ -cp_z & 0 & (E + mc^2) & 0 \\ 0 & cp_z & 0 & (E + mc^2) \end{vmatrix} \quad (3.18)$$

vanishes, that is, if $(E^2 - m^2c^4 - c^2p_z^2)^2 = 0$, or

$$E = \pm\sqrt{c^2p_z^2 + m^2c^4}\,.^2 \quad (3.19)$$

We therefore obtain the reasonable result that the system of equations (3.17) can be solved only on the condition that the relativistic energy law holds.

We shall now consider only the positive root in (3.19) (electrons) and not the negative energy states (positrons). Under the condition (3.19), the determinant (3.18) is of rank 2 (i.e., all 3 × 3 minor determinants vanish). This means that we obtain two linearly independent solutions from (3.17). We can write these as

$$a_1 = 1, \quad a_2 = 0, \quad a_3 = \frac{cp_z}{E + mc^2}, \quad a_4 = 0 \quad (3.20)$$

and

$$a_1 = 0, \quad a_2 = 1, \quad a_3 = 0, \quad a_4 = \frac{-cp_z}{E + mc^2}, \quad (3.21)$$

as can easily be verified by substituting into (3.17). By linearly combining these independent solutions, one obtains the general solution, so that we find for the general form of the plane wave

$$\left[A \begin{pmatrix} 1 \\ 0 \\ \frac{cp_z}{E + mc^2} \\ 0 \end{pmatrix} + B \begin{pmatrix} 0 \\ 1 \\ 0 \\ \frac{-cp_z}{E + mc^2} \end{pmatrix} \right] e^{i(kz - \omega t)}, \quad (3.22)$$

where A and B are constants.

---

[2] In this section E includes the rest energy, whereas we usually denote by E the energy without rest energy as is customary in low-energy physics.

Let us now investigate which spin states are described by the solutions obtained. For this we must first find the form of the spin operator in the Dirac theory. We start from the fact that the orbital angular momentum operator $\boldsymbol{l} = \boldsymbol{r} \times \boldsymbol{p}$ does not commute with the Hamiltonian operator of the Dirac theory. This means that $\boldsymbol{l}$ is not a constant of the motion, since for every operator Q that is not explicitly time dependent one has the relation

$$\frac{dQ}{dt} = \frac{1}{i\hbar}[Q, H] = \frac{1}{i\hbar}(QH - HQ).$$

This is different from what one would expect from classical physics or non-relativistic quantum mechanics where, for a central field, $\boldsymbol{l}$ is a constant of the motion.

We will show for the component l_x that $\boldsymbol{l}$ does not commute with the Hamiltonian

$$H = c\boldsymbol{\alpha} \cdot \boldsymbol{p} + \beta mc^2 + V(r): \qquad (3.23)$$

One has

$$[l_x, H] = c\left[(yp_z - zp_y)\left(\sum_\mu \alpha_\mu p_\mu + \beta mc + \frac{V(r)}{c}\right) \right.$$
$$\left. - \left(\sum_\mu \alpha_\mu p_\mu + \beta mc + \frac{V(r)}{c}\right)(yp_z - zp_y)\right].$$

The terms containing $V(r)$ cancel each other since $\boldsymbol{l}$ commutes with a spherically symmetric potential as one can recall from elementary quantum mechanics or easily check.

As the matrices α_μ and β contain only constants, they commute with coordinates and derivatives with respect to coordinates; thus

$$[l_x, H] = c\{\sum_\mu \alpha_\mu (yp_z - zp_y)p_\mu - \sum_\mu \alpha_\mu p_\mu (yp_z - zp_y)\}.$$

If one further observes that the space coordinates and likewise the momentum coordinates commute with each other and that one has

$$yp_y - p_y y = i\hbar, \quad xp_y - p_y x = 0, \quad \text{etc.}$$

then it follows that

$$[l_x, H] = -i\hbar c(\alpha_z p_y - \alpha_y p_z). \tag{3.24}$$

Corresponding results are obtained for the components l_y and l_z.

A theory which violates conservation of angular momentum for central forces is not satisfactory. Thus we must find an operator whose commutator with H is the negative of $[l, H]$; the sum of this operator and l would then commute with H (i.e., represent a constant of the motion) and could thus be conceived of as the total angular momentum operator. Such an operator can indeed be found; it is

$$s = \frac{\hbar}{2}\sigma$$

with

$$\sigma_x = \begin{pmatrix} 0 & 1 & 0 & 0 \\ 1 & 0 & 0 & 0 \\ 0 & 0 & 0 & 1 \\ 0 & 0 & 1 & 0 \end{pmatrix}, \quad \sigma_y = \begin{pmatrix} 0 & -i & 0 & 0 \\ i & 0 & 0 & 0 \\ 0 & 0 & 0 & -i \\ 0 & 0 & i & 0 \end{pmatrix},$$

$$\sigma_z = \begin{pmatrix} 1 & 0 & 0 & 0 \\ 0 & -1 & 0 & 0 \\ 0 & 0 & 1 & 0 \\ 0 & 0 & 0 & -1 \end{pmatrix} \tag{3.25}$$

which is a generalization of the Pauli spin operator (2.2). One can easily see (cf. Problem 3.1) that

$$[s, H] = -[l, H] \tag{3.26}$$

and thus

$$[(l + s), H] = 0. \tag{3.27}$$

Since $l + (\hbar/2)\sigma$ can be interpreted as the total angular momentum operator it appears obvious that $(\hbar/2)\sigma$ is the spin operator. (This argument will later be strengthened).

We now come back to the question of which spin states are represented by our solutions of the Dirac equation. It will be shown that (3.20) and (3.21) represent electron waves with spin parallel and antiparallel to the direction of propagation.

The first solution satisfies the equation[3]

$$\sigma_z \begin{pmatrix} 1 \\ 0 \\ \dfrac{cp_z}{E+mc^2} \\ 0 \end{pmatrix} e^{ikz} = 1 \cdot \begin{pmatrix} 1 \\ 0 \\ \dfrac{cp_z}{E+mc^2} \\ 0 \end{pmatrix} e^{ikz}, \qquad (3.28)$$

that is, the wave function given in (3.20) is an eigenfunction of $s_z = (\hbar/2)\sigma_z$ with the eigenvalue $+(\hbar/2)$. Similarly, because

$$\sigma_z \begin{pmatrix} 0 \\ 1 \\ 0 \\ \dfrac{-cp_z}{E+mc^2} \end{pmatrix} e^{ikz} = \begin{pmatrix} 0 \\ -1 \\ 0 \\ \dfrac{cp_z}{E+mc^2} \end{pmatrix} e^{ikz} = -1 \cdot \begin{pmatrix} 0 \\ 1 \\ 0 \\ \dfrac{-cp_z}{E+mc^2} \end{pmatrix} e^{ikz} \quad (3.29)$$

the solution (3.21) is an eigenfunction of s_z with eigenvalue $-(\hbar/2)$.

Contrary to the non-relativistic case, an eigenfunction of σ_x cannot be constructed now by the superposition of the eigenfunctions of σ_z with the eigenvalues ± 1. It can immediately be seen that the wave function formed by this superposition is not an eigenstate of σ_x:

$$\begin{pmatrix} 0 & 1 & 0 & 0 \\ 1 & 0 & 0 & 0 \\ 0 & 0 & 0 & 1 \\ 0 & 0 & 1 & 0 \end{pmatrix} \begin{pmatrix} 1 \\ 1 \\ \dfrac{cp_z}{E+mc^2} \\ \dfrac{-cp_z}{E+mc^2} \end{pmatrix} e^{ikz} = \begin{pmatrix} 1 \\ 1 \\ \dfrac{-cp_z}{E+mc^2} \\ \dfrac{cp_z}{E+mc^2} \end{pmatrix} e^{ikz}. \qquad (3.30)$$

This is not an eigenvalue equation, except for the special case $p_z = 0$.

This result was to be expected when one considers that it is not the spin but the total angular momentum that is a constant of the motion. Only if l or particular components l_μ vanish, are s or the corresponding s_μ constant. This is why for our plane wave in the z direction, for which $l_z = 0$ (but l_x, $l_y \neq 0$), one can find eigenvalues only of s_z. The components of orbital angular momentum l_x and l_y vanish only if $p = 0$. In this limiting case, eigenvalues of s_x and s_y will exist, as (3.30) shows for s_x.

---

[3] We neglect the irrelevant factor $\exp(-i\omega t)$.

3.1 The Dirac Equation and Its Interpretation 29

Hence it can be seen that in the relativistic case it is only in the rest frame of the electron that one can speak of a transverse spin direction of the plane wave (i.e., spins perpendicular to the direction of propagation). Only in the rest frame can one assign an eigenvalue to the spin operator in an arbitrary direction ϑ, φ. The spin part of the appropriate eigenfunction in the rest frame is, from (3.22),

$$\begin{pmatrix} A \\ B \\ 0 \\ 0 \end{pmatrix}, \tag{3.31}$$

so that in the limiting case where the momentum vanishes only two of the spinor components are different from zero. By referring to the results of Section 2.2 for two-component spinors, we see that the components of the spin function for the direction ϑ, φ which is defined in the rest frame are connected by the relation

$$\frac{B}{A} = \tan\frac{\vartheta}{2} e^{i\varphi} \tag{3.32}$$

[cf. (2.8)].

Since an arbitrary spin direction can be defined only in the rest frame, a definition of the polarization also makes sense only in the rest frame. Except for this restriction, we define the polarization as the expectation value of the spin operator, exactly as before.

To show what happens if one does calculate the polarization in the laboratory system instead of the rest system, we take, for example, the state

$$\chi^{(x)} = \begin{pmatrix} 1 \\ 1 \\ \dfrac{cp_z}{E + mc^2} \\ \dfrac{-cp_z}{E + mc^2} \end{pmatrix} e^{ikz}.$$

We have already seen from (3.30) that $\chi^{(x)}$ is not an eigensolution of σ_x in the laboratory system, but only in the rest frame. We calculate

$$P_x = \frac{\langle \chi^{(x)} | \sigma_x | \chi^{(x)} \rangle}{\langle \chi^{(x)} | \chi^{(x)} \rangle} \tag{3.33}$$

and find, using (3.30), that

$$P_x = \frac{2 \cdot \left(1 - \dfrac{c^2 p_z^2}{(E + mc^2)^2}\right)}{2 \cdot \left(1 + \dfrac{c^2 p_z^2}{(E + mc^2)^2}\right)}.$$

As $E = m\gamma c^2$ (with $\gamma = 1/(\sqrt{1 - \beta^2})$ and

$$c^2 p_z^2 = E^2 - m^2 c^4 = m^2 c^4 (\gamma^2 - 1)$$

we get

$$P_x = \frac{m^2 c^4 (\gamma + 1)^2 - m^2 c^4 (\gamma^2 - 1)}{m^2 c^4 (\gamma + 1)^2 + m^2 c^4 (\gamma^2 - 1)} = \frac{2\gamma + 2}{2\gamma^2 + 2\gamma}$$

$$= \frac{1}{\gamma} = \sqrt{1 - \beta^2}. \qquad (3.34)$$

Thus the polarization depends on the reference system; for an unambiguous definition it is therefore practical to refer it to the rest frame.

It is also possible to make "Lorentz invariant" (more precisely: covariant) definitions of the polarization, where it is not necessary to refer to the rest frame [3.2]. For our purposes, however, the definition as the expectation value of the spin operator in the rest frame suffices.

To conclude, we summarize what the Dirac equation describes:
a) Relativistic electrons (although it has not been shown here that the theory is Lorentz invariant, it appears plausible as it started from the relativistic energy equation),
b) the spin 1/2 of electrons,
c) the magnetic moment $\varepsilon\hbar/2mc$ of electrons,
d) spin-orbit coupling.

The last two points have not yet been shown here and their derivations will not be given in full. We will only outline the method of the somewhat tedious calculation.

If we consider electrons in external fields, we must again substitute $\boldsymbol{p} - (\varepsilon/c)\boldsymbol{A}$ and $H - \varepsilon\phi$ for $\boldsymbol{p}$ and H. The Dirac equation is then, see (3.12),

$$\left[H - \varepsilon\phi - c\boldsymbol{\alpha} \cdot \left(\boldsymbol{p} - \frac{\varepsilon}{c}\boldsymbol{A}\right) - \beta mc^2\right]\psi = 0. \qquad (3.35)$$

In order to be able to compare it with the Schrödinger equation, we reduce

(3.35) to the non-linearized form of (3.10), writing

$$\left[H - \varepsilon\phi - c\boldsymbol{\alpha}\cdot\left(\boldsymbol{p} - \frac{\varepsilon}{c}\boldsymbol{A}\right) - \beta mc^2 \right]$$
$$\times \left[H - \varepsilon\phi + c\boldsymbol{\alpha}\cdot\left(\boldsymbol{p} - \frac{\varepsilon}{c}\boldsymbol{A}\right) + \beta mc^2 \right]\psi = 0. \quad (3.36)$$

By multiplying out and making the approximation that the kinetic and potential energies are small compared with the rest energy mc^2 so that two components of the spin function can be neglected, one obtains

$$\left\{ \frac{1}{2m}\left(\boldsymbol{p} - \frac{\varepsilon}{c}\boldsymbol{A}\right)^2 + \varepsilon\phi - \frac{\varepsilon\hbar}{2mc}\boldsymbol{\sigma}\cdot\boldsymbol{B} + i\frac{\varepsilon\hbar}{4m^2c^2}\boldsymbol{E}\cdot\boldsymbol{p} \right.$$
$$\left. - \frac{\varepsilon\hbar}{4m^2c^2}\boldsymbol{\sigma}\cdot[\boldsymbol{E}\times\boldsymbol{p}] \right\}\psi = W\psi \quad (3.37)$$

when $W + mc^2$ is the total energy.[4]

The first two terms on the left are identical with those of the Schrödinger equation for external fields. The third term corresponds to the interaction energy $-\boldsymbol{\mu}\cdot\boldsymbol{B}$ between a magnetic dipole, whose moment is represented by the operator $\boldsymbol{\mu} = (\varepsilon\hbar/2mc)\boldsymbol{\sigma} = (\varepsilon/mc)\boldsymbol{s}$, and the external magnetic field. The fact that $(\varepsilon\hbar/2mc)\boldsymbol{\sigma}$ appears here as the operator of the magnetic moment, is a further reason for taking $(\hbar/2)\boldsymbol{\sigma}$ as the operator of the spin which is connected with this moment.

The fourth term is a relativistic correction to the energy and has no classical analogue. The meaning of the last term can again be illustrated. It describes the spin-orbit coupling. Since according to Maxwell's electrodynamics the vectors of the electromagnetic field are dependent on the reference system, an observer on an electron moving with velocity $\boldsymbol{v}$ relative to an electric field $\boldsymbol{E}$ finds a magnetic field[5] $\boldsymbol{B} = -c^{-1}\boldsymbol{v}\times\boldsymbol{E} = (mc)^{-1}[\boldsymbol{E}\times\boldsymbol{p}]$ (terms of the order $(v/c)^2$ are neglected). Thus, in its rest frame, an electron moving relative to the electric field of an atomic nucleus experiences this magnetic field, which affects its spin. The energy of the electron, due to its magnetic moment $\boldsymbol{\mu}$, in this field is

$$-\boldsymbol{\mu}\cdot\boldsymbol{B} = -\frac{\varepsilon}{mc}\boldsymbol{s}\cdot\boldsymbol{B} = -\frac{\varepsilon}{m^2c^2}\boldsymbol{s}\cdot[\boldsymbol{E}\times\boldsymbol{p}]. \quad (3.38)$$

Hence an additional energy term is obtained in the classical Hamiltonian

---

[4] For a more detailed discussion of this equation and the approximations made cf. [3.3 or 3.4].
[5] For the derivation of the formula refer to a textbook on electrodynamics.

3. Polarization Effects in Electron Scattering from Unpolarized Targets

function. If we substitute the spin operator $(\hbar/2)\boldsymbol{\sigma}$ for $\mathbf{s}$, we obtain the fifth term in the Hamiltonian operator (3.37) except for the factor 2. This factor is missing because our interpretation was too rough: We have not taken into account that in changing the frame of reference, the time transformation changes the precession frequency of the electron spin in the magnetic field (Thomas precession).

The term

$$-\frac{\varepsilon\hbar}{4m^2c^2}\boldsymbol{\sigma}\cdot[\mathbf{E}\times\mathbf{p}] \tag{3.39}$$

is called the spin-orbit energy, as it arises from the interaction of the spin with the magnetic field produced by the orbital motion of the electron, as we have illustrated. If the motion takes place in a central field of potential energy $V(r)$ where $\mathbf{E} = -\varepsilon^{-1}(dV/dr)(\mathbf{r}/r)$, we obtain from (3.39)

$$-\frac{\varepsilon}{2m^2c^2}\mathbf{s}\cdot\left[-\frac{1}{\varepsilon}\frac{dV}{dr}\frac{\mathbf{r}}{r}\times\mathbf{p}\right] = \frac{1}{2m^2c^2}\frac{1}{r}\frac{dV}{dr}(\mathbf{s}\cdot\mathbf{l}). \tag{3.40}$$

Thus the spin-orbit energy, in the case of the Coulomb potential, decreases more quickly with increasing distance than does the Coulomb energy itself and can therefore be neglected at fairly large distances from the nucleus.

Before we turn to the treatment of the scattering problem by the Dirac equation, it should be emphasized that spin, magnetic moment, and spin-orbit coupling, which are very important in the following sections, were not introduced by making additional assumptions. They follow automatically from the first principles from which Dirac's derivation started.

Problem 3.1: Show the validity of the formula

$$\frac{\hbar}{2}[\boldsymbol{\sigma}, H] = -[\mathbf{l}, H]$$

taking the x components as an example.

Solution: From (3.24) one has $[l_x, H] = -i\hbar c(\alpha_z p_y - \alpha_y p_z)$; thus only $[\sigma_x, H]$ remains to be calculated. As $\sigma_x V(r) = V(r)\sigma_x$ it follows that

$$[\sigma_x, H] = c[\sigma_x(\textstyle\sum_\mu \alpha_\mu p_\mu + \beta mc) - (\textstyle\sum_\mu \alpha_\mu p_\mu + \beta mc)\sigma_x].$$

One sees immediately from the matrices β and α_x that $\sigma_x\beta = \beta\sigma_x$ and $\sigma_x\alpha_x = \alpha_x\sigma_x$. Thus

$$[\sigma_x, H] = c[\sigma_x(\alpha_y p_y + \alpha_z p_z) - (\alpha_y p_y + \alpha_z p_z)\sigma_x].$$

Due to the relations

$$\sigma_x\alpha_y = \begin{pmatrix} 0 & 0 & i & 0 \\ 0 & 0 & 0 & -i \\ i & 0 & 0 & 0 \\ 0 & -i & 0 & 0 \end{pmatrix} = i\alpha_z, \quad \alpha_y\sigma_x = \begin{pmatrix} 0 & 0 & -i & 0 \\ 0 & 0 & 0 & i \\ -i & 0 & 0 & 0 \\ 0 & i & 0 & 0 \end{pmatrix} = -i\alpha_z,$$

$$\sigma_x\alpha_z = \begin{pmatrix} 0 & 0 & 0 & -1 \\ 0 & 0 & 1 & 0 \\ 0 & -1 & 0 & 0 \\ 1 & 0 & 0 & 0 \end{pmatrix} = -i\alpha_y, \quad \alpha_z\sigma_x = \begin{pmatrix} 0 & 0 & 0 & 1 \\ 0 & 0 & -1 & 0 \\ 0 & 1 & 0 & 0 \\ -1 & 0 & 0 & 0 \end{pmatrix} = i\alpha_y$$

and the fact that p commutes with σ one obtains

$$\frac{\hbar}{2}[\sigma_x, H] = i\hbar c(\alpha_z p_y - \alpha_y p_z) = -[l_x, H].$$

3.2 Calculation of the Differential Scattering Cross Section

The differential cross section for elastic scattering of an electron beam with arbitrary spin direction is calculated using the Dirac equation. The scattering cross section depends on the azimuthal angle ϕ; this means that there is generally no axial symmetry of the scattered intensity with respect to the incident direction. The asymmetry is described by the Sherman function.

We are now in a position to deal with the scattering of relativistic electrons with spin by a central field. The electrons will be taken as an incident plane wave in the z direction. In analogy to non-relativistic scattering theory, we look for solutions to the Dirac equation with the asymptotic form

$$\psi_\lambda \xrightarrow[r \to \infty]{} a_\lambda e^{ikz} + a'_\lambda(\theta, \phi) \frac{e^{ikr}}{r} \tag{3.41}$$

for the four components of the wave function ($\lambda = 1, \ldots, 4$). Generalizing the differential elastic scattering cross section obtained from the Schrödinger theory one finds

$$\frac{d\sigma}{d\Omega}(\theta, \phi) \equiv \sigma(\theta, \phi) = \frac{\sum_{\lambda=1}^{4} |a'_\lambda(\theta, \phi)|^2}{\sum_{\lambda=1}^{4} |a_\lambda|^2}. \tag{3.42}$$

This follows from the general definition of the differential cross section and from the fact that the current density can be written as $\rho v = \psi^\dagger \psi v$. If one uses normalized wave functions, the denominator in (3.42) is 1.

In order to simplify this expression, we use the fact that the a_λ are not independent of each other. This can be seen from the solution (3.22) for a plane wave with arbitrary spin direction which shows that

$$r \equiv \frac{|a_3|}{|a_1|} = \frac{|a_4|}{|a_2|} = \frac{cp_z}{E + mc^2}. \tag{3.43}$$

The same relation exists asymptotically between the a'_λ because at very large distances from the scattering center, the scattered spherical wave can be regarded as made up of plane waves proceeding in different directions from the center. Therefore we have

$$\sigma(\theta, \phi) = \frac{|a'_1|^2 + |a'_2|^2 + |a'_1|^2 r^2 + |a'_2|^2 r^2}{|a_1|^2 + |a_2|^2 + |a_1|^2 r^2 + |a_2|^2 r^2} = \frac{|a'_1|^2 + |a'_2|^2}{|a_1|^2 + |a_2|^2}. \quad (3.44)$$

In the following we consider the scattering of electron waves whose spins are oriented in the $+z$ or $-z$ direction. If we have the solutions to the scattering problem for these two basic states we can construct all other cases from them. By coherent superposition we obtain, for example, scattering of a beam with arbitrary spin direction and $P = 1$ (see Sect. 2.2); by incoherent superposition we obtain scattering of an unpolarized beam (see Sect. 2.3).

In the case in which the spin of the incident wave is in the $+z$ direction, its wave function, from (3.28), is

$$\begin{pmatrix} \psi_1 \\ \psi_2 \\ \vdots \end{pmatrix} = \begin{pmatrix} 1 \\ 0 \\ \vdots \end{pmatrix} e^{ikz}.$$

The "small" components ψ_3 and ψ_4, due to their dependence on the "large" components ψ_1 and ψ_2 [see 3.43], yield no additional information as is shown clearly by (3.44); they therefore need not be considered. To solve the scattering problem for this particular choice of incident wave one must look for solutions of the Dirac equation with the asymptotic form

$$\psi_1 \xrightarrow[r \to \infty]{} e^{ikz} + f_1(\theta, \phi) \frac{e^{ikr}}{r}$$

$$\psi_2 \xrightarrow[r \to \infty]{} 0 + g_1(\theta, \phi) \frac{e^{ikr}}{r}. \quad (3.45)$$

This takes account of the fact that the second component of the wave function is no longer necessarily zero after scattering as the spin may change its direction due to spin-orbit coupling. The electron "sees" in its rest frame the moving charge of the scattering center; that means it sees a current and thus a magnetic field, which acts upon its magnetic moment and may change its spin direction (see end of Sect. 3.1). This possibility is taken into account through the inclusion of g_1, which is therefore called the spin-flip amplitude.

3.2 Calculation of the Differential Scattering Cross Section

Analogously, for the incident wave with the other spin direction, we expect a solution with the asymptotic form

$$\psi_1 \xrightarrow[r \to \infty]{} 0 + g_2(\theta, \phi) \frac{e^{ikr}}{r}$$

$$\psi_2 \xrightarrow[r \to \infty]{} e^{ikz} + f_2(\theta, \phi) \frac{e^{ikr}}{r}.$$

(3.46)

As in the case of the Schrödinger equation, the scattering problem can be solved by using the method of partial waves in which one looks for particular solutions with specific angular momenta and constructs from them a general solution with the required boundary conditions. In our case, the procedure is the same, but the solution is considerably more complicated. This is because one does not (as in the Schrödinger theory) have only one differential equation for one function, but instead has a system of simultaneous differential equations. We shall not reproduce the calculation step by step—it can be found in MOTT and MASSEY [3.1]. We will only explain the essential ideas and emphasize the physical background.

Let us first consider the case in which the spin of the incident beam is in the $+z$ direction which we take as the quantization axis. By separating the variables in the Dirac equation for a central field, one obtains the particular solutions

$$\begin{pmatrix} \psi_1 \\ \psi_2 \end{pmatrix} = \begin{pmatrix} (l+1)G_l(r)P_l(\cos\theta) \\ -G_l(r)P_l^1(\cos\theta)e^{i\phi} \end{pmatrix}$$

and

(3.47)

$$\begin{pmatrix} \psi_1 \\ \psi_2 \end{pmatrix} = \begin{pmatrix} lG_{-l-1}(r)P_l(\cos\theta) \\ G_{-l-1}(r)P_l^1(\cos\theta)e^{i\phi} \end{pmatrix},$$

where $P_l^1(\theta)$ are associated Legendre functions; ϕ is the azimuthal angle. Thus one finds a *pair* of solutions just as in the case of the plane wave.

There the two solutions differed by the spin directions which they described. What we have obtained for a central potential is quite analogous, as will now be illustrated.

The functions $G_l(r)$ and $G_{-l-1}(r)$ are solutions of the two r-dependent ordinary differential equations which arise from separating the variables in the Dirac equation. The fact that here, contrary to the Schrödinger theory, not one but two radial differential equations appear can be explained by the spin-orbit coupling term. For this we consider once again the case of small velocities as at the end of Section 3.1. Then the operator for spin-orbit coupling is proportional to $r^{-1}(dV/dr)(\mathbf{l}\cdot\mathbf{s})$ [see (3.40)], and since $\mathbf{j}^2 = (\mathbf{l}+\mathbf{s})^2 = \mathbf{l}^2 + \mathbf{s}^2 + 2\mathbf{l}\cdot\mathbf{s}$, it is proportional to $r^{-1}(dV/dr)(\mathbf{j}^2 - \mathbf{l}^2 - \mathbf{s}^2)$. When this r-dependent operator is applied to the wave function it produces a term in the radial differential equation which is proportional to $r^{-1}(dV/dr)[j(j+1) - l(l+1) - s(s+1)]$. Since

$$[j(j+1) - l(l+1) - s(s+1)] = \begin{cases} l, & \text{if } j = l + \tfrac{1}{2} \\ -l-1, & \text{if } j = l - \tfrac{1}{2} \end{cases} \quad (3.48)$$

one obtains a different differential equation for each of the two spin orientations and thus two different solutions G_l and G_{-l-1}.

The physically important occurrence in (3.47) of a ϕ-dependent term which does not occur in the Schrödinger treatment of scattering can also be intuitively explained. Due to the conservation law for j, the angular momentum component m_j in the z direction must be $+(1/2)$ after the scattering, since before the scattering $m_s = +(1/2)$, $m_l = 0$, i.e., $m_j = +(1/2)$, according to the initial assumption. For a spin flip as described by ψ_2 the decrease of the z component of the spin must be compensated for by an increase of the orbital angular momentum component in this direction. Thus a non-zero expectation value for the z component of the orbital angular momentum must exist. Since the operator of this component is

$$l_z = -i\hbar \frac{\partial}{\partial \phi},$$

this is possible only if the solution contains a ϕ-dependent term.

As with the treatment of scattering in the Schrödinger theory, the solutions of the radial differential equations for potentials which decrease faster than r^{-1} have the asymptotic form

$$G_l \xrightarrow[r\to\infty]{} \frac{\sin[kr - (l\pi/2) + \eta_l]}{kr},$$

$$G_{-l-1} \xrightarrow[r\to\infty]{} \frac{[\sin kr - (l\pi/2) + \eta_{-l-1}]}{kr}. \quad (3.49)$$

In the important case of the r^{-1}-potential a logarithmic term must be included in the argument of the sine function; everything else remains valid.

Let us now construct from the particular solutions (3.47) (partial waves) a solution which has the required form (3.45). This can be done if they are combined as follows

$$\psi_1 = \sum_{l=0}^{\infty} \{(l+1) e^{i\eta_l} G_l + l e^{i\eta_{-l-1}} G_{-l-1}\} i^l P_l(\cos\theta)$$
$$\psi_2 = \sum_{l=1}^{\infty} \{-e^{i\eta_l} G_l + e^{i\eta_{-l-1}} G_{-l-1}\} i^l P_l^1(\cos\theta) e^{i\phi}. \tag{3.50}$$

One can easily check (see Problem 3.2) that the condition (3.45) is then fulfilled; one needs only to choose

$$f_1(\theta, \phi) = \frac{1}{2ik} \sum_{l=0}^{\infty} [(l+1)(e^{2i\eta_l} - 1) + l(e^{2i\eta_{-l-1}} - 1)] P_l(\cos\theta)$$

$$\equiv f(\theta) \tag{3.51}$$

and

$$g_1(\theta, \phi) = \frac{1}{2ik} \sum_{l=1}^{\infty} (-e^{2i\eta_l} + e^{2i\eta_{-l-1}}) P_l^1(\cos\theta) e^{i\phi} \equiv g(\theta) e^{i\phi}.$$

The solution of the scattering problem is thus reduced to a calculation of the scattering phases η. These depend, as in the non-relativistic case, on the energy of the incident electrons and on the scattering potential (scattering substance). Hence the scattering amplitudes f and g, apart from depending on the scattering angle, also depend on these two variables.

An analogous treatment for the spin of the incident wave in the $-z$ direction yields for f_2 and g_2 in (3.46)

$$f_2 = f_1 = f(\theta)$$
$$g_2 = -g_1 e^{-2i\phi} = -g(\theta) e^{-i\phi}, \tag{3.52}$$

with f_1 and g_1 from (3.51).

We can now easily treat the case of an incident wave with an arbitrary spin direction

$$A \begin{pmatrix} 1 \\ 0 \end{pmatrix} e^{ikz} + B \begin{pmatrix} 0 \\ 1 \end{pmatrix} e^{ikz} = \begin{pmatrix} A \\ B \end{pmatrix} e^{ikz}, \tag{3.53}$$

where A and B, according to (3.32), specify the spin direction in the rest frame. Using (3.45), (3.46), (3.52) and (3.53), one obtains by coherent superposition

$$\begin{pmatrix} a_1' \\ a_2' \end{pmatrix} \frac{e^{ikr}}{r} = A \begin{pmatrix} f_1 \\ g_1 \end{pmatrix} \frac{e^{ikr}}{r} + B \begin{pmatrix} g_2 \\ f_2 \end{pmatrix} \frac{e^{ikr}}{r} = \begin{pmatrix} Af - Bge^{-i\phi} \\ Bf + Age^{i\phi} \end{pmatrix} \frac{e^{ikr}}{r} \tag{3.54}$$

for the asymptotic form of the scattered wave. Thus from (3.44) the differential cross section is

$$\sigma(\theta, \phi) = \frac{|a_1'|^2 + |a_2'|^2}{|A|^2 + |B|^2}$$

$$= |f|^2 + |g|^2 + \frac{-AB^* e^{i\phi} + A^* B e^{-i\phi}}{|A|^2 + |B|^2} (fg^* - f^*g), \quad (3.55)$$

which shows that for a polarized primary beam the scattering intensity generally depends on ϕ. By substituting

$$S(\theta) = i \frac{fg^* - f^*g}{|f|^2 + |g|^2} \quad (3.56)$$

(Sherman function; it is real as its numerator is the difference of two conjugate complex functions), it follows that

$$\sigma(\theta, \phi) = (|f|^2 + |g|^2) \left[1 + S(\theta) \frac{-AB^* e^{i\phi} + A^* B e^{-i\phi}}{i(|A|^2 + |B|^2)} \right]. \quad (3.57)$$

Example 3.1:
$A = 1$ and $B = 1$, i.e., transverse polarization, since [cf. Eqs. (3.32) to (3.34)] the primary beam is totally polarized in the x direction. From (3.57) one has $\sigma(\theta, \phi) = (|f|^2 + |g|^2) [1 - S(\theta) \sin \phi]$. The ϕ-dependence of the cross section is not surprising since the primary beam is not axially symmetric with respect to the direction of propagation. As shown in Figs. 3.1 and 3.2, the scattering asymmetry is maximum when the scattering

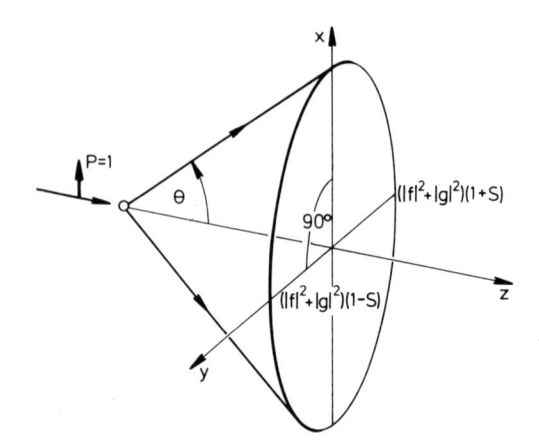

Fig. 3.1. Left-right scattering asymmetry of a beam totally polarized in the x direction

3.2 Calculation of the Differential Scattering Cross Section

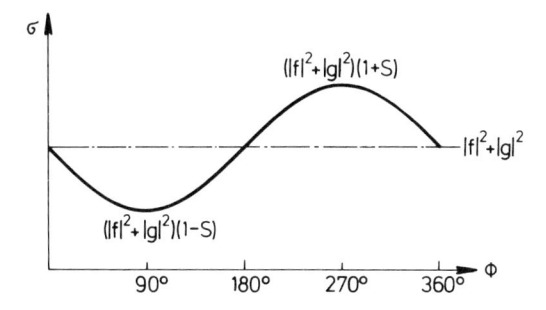

Fig. 3.2. Dependence of the differential cross section on the azimuthal angle ϕ (for $S > 0$; see Sect. 3.5)

plane (plane described by primary beam and direction of observation) is perpendicular to P, that is, when $\phi = 90°$ and $270°$. This "left-right" asymmetry of the scattering is used to measure the polarization of electron beams ("Mott[6] detector"). The scattering intensities "up" and "down" ($\phi = 0°$ and $180°$) are equal.

Example 3.2:

$A = 1$, $B = 0$ or $A = 0$, $B = 1$, i.e., longitudinal polarization. These are our basic functions with spin parallel or antiparallel to the incident direction. In this case, the ϕ-dependent part of the scattering cross section disappears according to (3.57). This is to be expected, as here—unlike the first example—the incident beam is axially symmetric.

Problem 3.2: Prove that the solution (3.50) has the required asymptotic form (3.45), if (3.49) and (3.51) are fulfilled.

Solution: Using the relations (3.49) one obtains from (3.50)

$$\psi_1 \xrightarrow[r \to \infty]{} \sum_{l=0}^{\infty} \left[(l+1) \exp(i\eta_l) \frac{\exp\left[i\left(kr - \frac{l\pi}{2} + \eta_l\right)\right] - \exp\left[-i\left(kr - \frac{l\pi}{2} + \eta_l\right)\right]}{2ikr} \right.$$

$$\left. + l \exp(i\eta_{-l-1}) \frac{\exp\left[i\left(kr - \frac{l\pi}{2} + \eta_{-l-1}\right)\right] - \exp\left[-i\left(kr - \frac{l\pi}{2} + \eta_{-l-1}\right)\right]}{2ikr} \right] i^l P_l$$

$$= \frac{1}{2ik} \sum_{l=0}^{\infty} \left\{ [(l+1) \exp(2i\eta_l) + l \exp(2i\eta_{-l-1})] \frac{e^{ikr}}{r} \right.$$

$$\left. - \left[(l+1) \frac{\exp\left[-i\left(kr - \frac{l\pi}{2}\right)\right]}{r} + l \frac{\exp\left[-i\left(kr - \frac{l\pi}{2}\right)\right]}{r} \right] i^l \right\} P_l.$$

Adding

$$\left\{ -[(l+1) + l] \frac{e^{ikr}}{r} + i^l(2l+1) \frac{\exp\left[i\left(kr - \frac{l\pi}{2}\right)\right]}{r} \right\} P_l = 0$$

---

[6] The relativistic theory of electron scattering dealt with in this section originated with N. F. MOTT (see [3.1]; further references are given there).

yields

$$\psi_1 \xrightarrow[r\to\infty]{} \frac{1}{2ik} \sum_{l=0}^{\infty} \Big\{ [(l+1)\{\exp(2i\eta_l) - 1\} + l\{\exp(2i\eta_{-l-1}) - 1\}] P_l \frac{e^{ikr}}{r}$$

$$+ (2l+1)i^l \cdot \frac{\exp\left[i\left(kr - \frac{l\pi}{2}\right)\right] - \exp\left[-i\left(kr - \frac{l\pi}{2}\right)\right]}{r} P_l \Big\}$$

$$= f_1(\theta, \phi) \frac{e^{ikr}}{r} + \sum_{l=0}^{\infty} (2l+1)i^l \frac{\sin\left(kr - \frac{l\pi}{2}\right)}{kr} P_l;$$

hence

$$\psi_1 \xrightarrow[r\to\infty]{} f_1(\theta, \phi) \frac{e^{ikr}}{r} + e^{ikz},$$

where we have used the asymptotic expansion for exp (*ikz*) which can be found in the treatment of scattering with the Schrödinger equation in textbooks on quantum mechanics.

Similarly one obtains

$$\psi_2 \xrightarrow[r\to\infty]{} \frac{1}{2ik} \sum_{l=1}^{\infty} \Big[-\exp(i\eta_l) \frac{\exp\left[i\left(kr - \frac{l\pi}{2} + \eta_l\right)\right] - \exp\left[-i\left(kr - \frac{l\pi}{2} + \eta_l\right)\right]}{r}$$

$$+ \exp(i\eta_{-l-1}) \frac{\exp\left[i\left(kr - \frac{l\pi}{2} + \eta_{-l-1}\right)\right] - \exp\left[-i\left(kr - \frac{l\pi}{2} + \eta_{-l-1}\right)\right]}{r} \Big] i^l P_l^1 e^{i\phi}$$

$$= \frac{1}{2ik} \sum_{l=1}^{\infty} \frac{e^{ikr}}{r} [-\exp(2i\eta_l) + \exp(2i\eta_{-l-1})] P_l^1 e^{i\phi}$$

$$= g_1(\theta, \phi) \frac{e^{ikr}}{r}.$$

3.3 The Role of Spin Polarization in Scattering

The following will be shown with the use of density matrix formalism: Only the component of the polarization vector which is perpendicular to the scattering plane contributes to the scattering asymmetry. An originally unpolarized electron beam has, after scattering, a polarization of magnitude $S(\theta)$ [$S(\theta)$ = Sherman function] perpendicular to the scattering plane. The direction and amplitude of the polarization vector of an arbitrarily polarized primary beam are usually changed by scattering. However, a totally polarized beam remains totally polarized, and a beam polarized perpendicular to the scattering plane retains its direction of polarization. Double scattering experiments are suitable for determining the Sherman function (except for the sign).

3.3.1 Polarization Dependence of the Cross Section

We will now write the differential scattering cross section in a form which shows the influence of the polarization more clearly. For this, we recall the equations (3.53) and (3.54), which show that the spinor $\chi = \begin{pmatrix} A \\ B \end{pmatrix}$ of a pure

3.3 The Role of Spin Polarization in Scattering

initial state is transformed by the scattering process to the spinor

$$\chi' = \begin{pmatrix} Af - Bge^{-i\phi} \\ Bf + Age^{i\phi} \end{pmatrix} \qquad (3.58)$$

of the final state. This can be mathematically represented as transformation

$$\chi' = \begin{pmatrix} f & -ge^{-i\phi} \\ ge^{i\phi} & f \end{pmatrix} \begin{pmatrix} A \\ B \end{pmatrix} = S\chi \qquad (3.59)$$

by means of a matrix, the scattering matrix S for the spin.

The density matrix ρ' for the scattered state, which we again abbreviate as $\chi' = \begin{pmatrix} a'_1 \\ a'_2 \end{pmatrix}$, is [cf. (2.22)]

$$\rho' = \begin{pmatrix} |a'_1|^2 & a'_1 a'^*_2 \\ a'^*_1 a'_2 & |a'_2|^2 \end{pmatrix} = \begin{pmatrix} a'_1 \\ a'_2 \end{pmatrix}(a'^*_1, a'^*_2) = \chi'\chi'\dagger = S\chi\chi\dagger S\dagger. \qquad (3.60)$$

Since $\chi\chi\dagger = \rho$ (density matrix of the unscattered state), it follows that[7]

$$\rho' = S\rho S\dagger = \tfrac{1}{2} S(1 + \mathbf{P}\cdot\boldsymbol{\sigma})S\dagger \, \text{tr}\, \rho. \qquad (3.61)$$

Equation (2.23) for unnormalized wave functions has been used here because we did not consider normalization in our treatment of scattering.

If we don't have a pure initial state but a statistical mixture of spin states, i.e., a partially polarized primary beam, (3.61) is valid as it stands; ρ and ρ' are then the respective density matrices of the mixed initial and final states (see Problem 3.3).

Equation (3.61) can be used to directly write the dependence of the differential cross section on the polarization $\mathbf{P}$ of the incident beam. According to (3.54) and (3.55), and from (3.60), one has

$$\sigma(\theta, \phi) = \frac{|Af - Bge^{-i\phi}|^2 + |Bf + Age^{i\phi}|^2}{|A|^2 + |B|^2} = \frac{\text{tr}\, \rho'}{\text{tr}\, \rho}. \qquad (3.62)$$

Therefore with (3.61) the dependence of the differential cross section on the polarization of the primary beam is

$$\sigma(\theta, \phi) = \tfrac{1}{2} \, \text{tr}\, S(1 + \mathbf{P}\cdot\boldsymbol{\sigma})S\dagger. \qquad (3.63)$$

---

[7] Recall that we defined the polarization in the rest frame of the electrons. According to (3.31), we then need only two spinor components, just as in the calculation of the scattering cross section. We can therefore use the density matrix formalism which was developed in Section 2.3 for two-component spinors.

3. Polarization Effects in Electron Scattering from Unpolarized Targets

To evaluate this one must form the trace of the product

$$\begin{pmatrix} f & -ge^{-i\phi} \\ ge^{i\phi} & f \end{pmatrix} \begin{pmatrix} 1+P_z & P_x-iP_y \\ P_x+iP_y & 1-P_z \end{pmatrix} \begin{pmatrix} f^* & g^*e^{-i\phi} \\ -g^*e^{i\phi} & f^* \end{pmatrix}.$$

The simple calculation yields

$$\sigma(\theta,\phi) = (|f|^2+|g|^2)\left\{1 - \frac{S(\theta)}{2i}[e^{i\phi}(P_x-iP_y) - e^{-i\phi}(P_x+iP_y)]\right\}.$$

(3.64)

[$S(\theta)$ is the Sherman function (3.56) and should not be confused with the scattering matrix S.] The differential cross section is thus independent of the longitudinal polarization component P_z.

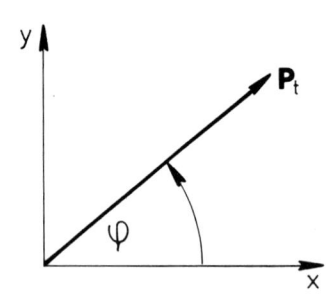

Fig. 3.3. Transverse polarization component

With $P_t \exp(\pm i\varphi) = P_x \pm iP_y$, where P_t is the magnitude of the transverse polarization component P_t (cf. Fig. 3.3), and

$$|f|^2 + |g|^2 = I(\theta) \tag{3.65}$$

one has

$$\sigma(\theta,\phi) = I(\theta)\left\{1 - \frac{S(\theta)P_t}{2i}[e^{i(\phi-\varphi)} - e^{-i(\phi-\varphi)}]\right\}$$
$$= I(\theta)\{1 - P_t S(\theta)\sin(\phi-\varphi)\}. \tag{3.66}$$

This shows that for an electron beam which has no transverse polarization the differential cross section is independent of the azimuthal angle ϕ and simply has the value $I(\theta) = |f|^2 + |g|^2$.

Equation (3.66) can be further simplified if one defines the direction of the transverse polarization component as the x direction (see Fig. 3.4).

Fig. 3.4. Scattering of a polarized beam

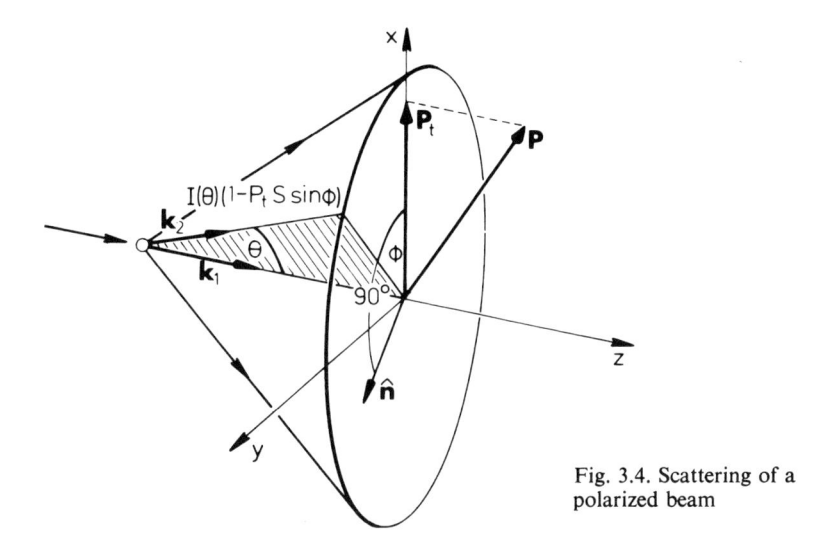

Then it follows that the differential cross section for a primary beam with arbitrary polarization is

$$\sigma(\theta, \phi) = I(\theta)\{1 - P_t S(\theta) \sin \phi\}. \tag{3.67}$$

Frequently this formula is written using the unit vector perpendicular to the scattering plane

$$\hat{n} = \frac{k_1 \times k_2}{|k_1 \times k_2|} \tag{3.68}$$

(k_1 and k_2 are, except for the factor $\hbar$, the electron momenta before and after scattering). Since in our suitably chosen coordinate system we have $P = (P_x, 0, P_z)$, $P_x = P_t$, and

$$\hat{n} = (-\sin \phi, \cos \phi, 0) \tag{3.69}$$

(cf. Fig. 3.4), we obtain $-P_t \sin \phi = P \cdot \hat{n}$ and thus from (3.67)

$$\sigma(\theta, \phi) = I(\theta)[1 + S(\theta) P \cdot \hat{n}]. \tag{3.70}$$

This formula is independent of the choice of coordinate system as the scalar product is invariant under coordinate transformations.

This is the basic equation for the measurement of electron polarization by "Mott scattering". An essential feature of this formula is that only the component of the polarization vector perpendicular to the scattering plane contributes to the scattering asymmetry; components parallel to the scattering plane make no contribution (see also Problem 3.5).

Problem 3.3: Relation (3.61) has been derived for the scattering of a pure spin state. Show that it is also valid for the scattering of a mixture of spin states, when the density matrices of the mixed states are defined by (2.22).

Solution: One has
$$\rho' = \Sigma_n \rho'^{(n)} = \Sigma_n S\rho^{(n)}S\dagger$$
$$= \tfrac{1}{2}S\{\Sigma_n (1 + \boldsymbol{P}^{(n)} \cdot \boldsymbol{\sigma}) \operatorname{tr} \rho^{(n)}\}S\dagger.$$
As $\Sigma_n \operatorname{tr} \rho^{(n)} = \operatorname{tr} \rho$ and $\Sigma_n \boldsymbol{P}^{(n)} \operatorname{tr} \rho^{(n)} = \boldsymbol{P} \operatorname{tr} \rho$ [see (2.20)], it follows
$$\rho' = \tfrac{1}{2}S(1 + \boldsymbol{P} \cdot \boldsymbol{\sigma})S\dagger \operatorname{tr} \rho.$$

3.3.2 Polarization of an Electron Beam by Scattering

The density-matrix formalism readily produces the remarkable fact that an initially unpolarized electron beam is polarized by the scattering process: From (2.21) one has

$$\boldsymbol{P}' = \frac{\operatorname{tr} \rho' \boldsymbol{\sigma}}{\operatorname{tr} \rho'}.$$

If the primary beam is unpolarized one obtains from (3.61), as $\boldsymbol{P} = 0$,

$$\rho' = \tfrac{1}{2}S(1 + \boldsymbol{P} \cdot \boldsymbol{\sigma})S\dagger \operatorname{tr} \rho = \tfrac{1}{2}SS\dagger \operatorname{tr} \rho \tag{3.71}$$

so that the polarization of the scattered beam is

$$\boldsymbol{P}' = \frac{1}{2} \frac{\operatorname{tr} SS\dagger \boldsymbol{\sigma}}{\operatorname{tr} \rho'} \operatorname{tr} \rho. \tag{3.72}$$

From (3.62) one has $\operatorname{tr} \rho'/\operatorname{tr} \rho = \sigma(\theta, \phi) = |f|^2 + |g|^2$, where (3.66) has been applied in the special case of an unpolarized primary beam. Furthermore, with $\boldsymbol{\sigma} = \Sigma_\mu \sigma_\mu \hat{\boldsymbol{e}}_\mu$ [where $\hat{\boldsymbol{e}}_\mu$ are unit vectors along the coordinate axes and the σ_μ are defined by (2.2)], one has

$$\tfrac{1}{2} \operatorname{tr} (SS\dagger \boldsymbol{\sigma}) = \tfrac{1}{2} \operatorname{tr} \begin{pmatrix} |f|^2 + |g|^2 & fg^*e^{-i\phi} - f^*ge^{-i\phi} \\ f^*ge^{i\phi} - fg^*e^{i\phi} & |f|^2 + |g|^2 \end{pmatrix}$$
$$\times \begin{pmatrix} \hat{\boldsymbol{e}}_z & \hat{\boldsymbol{e}}_x - i\hat{\boldsymbol{e}}_y \\ \hat{\boldsymbol{e}}_x + i\hat{\boldsymbol{e}}_y & -\hat{\boldsymbol{e}}_z \end{pmatrix}$$
$$= \tfrac{1}{2}[(|f|^2 + |g|^2)\hat{\boldsymbol{e}}_z + (fg^* - f^*g)(\hat{\boldsymbol{e}}_x + i\hat{\boldsymbol{e}}_y)e^{-i\phi}$$
$$- (fg^* - f^*g)(\hat{\boldsymbol{e}}_x - i\hat{\boldsymbol{e}}_y)e^{i\phi} - (|f|^2 + |g|^2)\hat{\boldsymbol{e}}_z]$$
$$= i(fg^* - f^*g)(-\sin \phi \cdot \hat{\boldsymbol{e}}_x + \cos \phi \cdot \hat{\boldsymbol{e}}_y).$$

Since from (3.69) $\hat{n} = (-\sin\phi, \cos\phi, 0)$, one finally obtains, by using the Sherman function (3.56),

$$P' = S(\theta)\hat{n}. \tag{3.73}$$

Thus, through scattering, an initially unpolarized beam acquires a polarization of magnitude $S(\theta)$ perpendicular to the scattering plane. This, together with the result of the last section, shows that the Sherman function describes two important features: The extent of the left-right asymmetry in the scattering of a polarized beam and the amount of polarization produced by scattering an unpolarized beam.

3.3.3 Behavior of the Polarization in Scattering

The problem just treated is a special case of the following: Given an incident electron beam with arbitrary polarization P, how is this polarization changed by scattering?

The polarization P' after scattering is

$$P' = \frac{\operatorname{tr} \rho'\sigma}{\operatorname{tr} \rho'} = \frac{\frac{1}{2}\operatorname{tr} S(1 + P\cdot\sigma)S^\dagger\sigma}{\frac{1}{2}\operatorname{tr} S(1 + P\cdot\sigma)S^\dagger}. \tag{3.74}$$

The denominator has already been evaluated [cf. (3.63) to (3.70)] with the result $(|f|^2 + |g|^2)[1 + S(\theta)P\cdot\hat{n}]$. The numerator is calculated in Problem 3.4. As a result, one obtains for the polarization after scattering

$$P' = \frac{[P\cdot\hat{n} + S(\theta)]\hat{n} + T(\theta)[P - (P\cdot\hat{n})\hat{n}] + U(\theta)[\hat{n} \times P]}{1 + P\cdot\hat{n}S(\theta)} \tag{3.75}$$

with

$$S(\theta) = i\frac{fg^* - f^*g}{|f|^2 + |g|^2}, \quad T(\theta) = \frac{|f|^2 - |g|^2}{|f|^2 + |g|^2}, \quad U(\theta) = \frac{fg^* + f^*g}{|f|^2 + |g|^2}. \tag{3.76}$$

If one resolves the polarization vector into components P_p parallel to the scattering plane and P_n perpendicular to it (i.e., parallel to $\hat{n}$, see Fig. 3.5), one has $P = P_n + P_p$. As $(P\cdot\hat{n})\hat{n} = P_n$ and $\hat{n} \times P_n = 0$ or $\hat{n} \times P = \hat{n} \times P_p$, (3.75) can also be written in the form

$$P' = \frac{[P_n + S(\theta)]\hat{n} + T(\theta)P_p + U(\theta)[\hat{n} \times P_p]}{1 + P_n S(\theta)}. \tag{3.77}$$

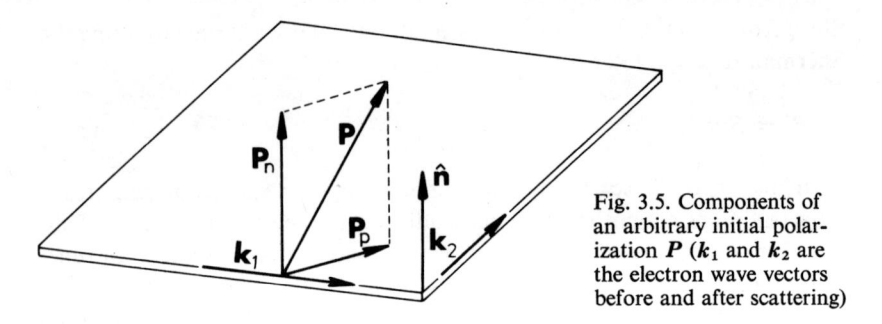

46   3. Polarization Effects in Electron Scattering from Unpolarized Targets

Fig. 3.5. Components of an arbitrary initial polarization P (k_1 and k_2 are the electron wave vectors before and after scattering)

In this formula only the essential components of the initial polarization appear. By use of the vector relation

$$a \times [b \times c] = (a \cdot c)b - (a \cdot b)c,$$

which in our case yields

$$\hat{n} \times [P \times \hat{n}] = P - (P \cdot \hat{n})\hat{n} = P_p,$$

(3.77) can be rearranged. Then one obtains the frequently used formula in which the complete vector P appears

$$P' = \frac{[P \cdot \hat{n} + S(\theta)]\hat{n} + T(\theta)\hat{n} \times [P \times \hat{n}] + U(\theta)[\hat{n} \times P]}{1 + P \cdot \hat{n} S(\theta)}. \qquad (3.78)$$

From (3.77) we easily see that the scattering process affects the polarization as follows: The (positive or negative) vector $S(\theta)\hat{n}$ is added to the component $P_n \hat{n}$ perpendicular to the scattering plane. The component parallel to the scattering plane is reduced from P_p to $T \cdot P_p$ ($|T| \leq 1$ by definition). The polarization vector is rotated out of its original plane (P_n, P_p) [identical to the plane ($\hat{n}$, P_p)] as there is an additional component, determined by $U(\theta)$, that is perpendicular to this plane. The change of the polarization vector due to the scattering is determined by the spin-flip amplitude g: If $g = 0$, then $T = 1$, $S = U = 0$, and according to (3.77) the polarization does not change.

The angle α, through which the polarization component P_p in the scattering plane is rotated, is given by (see Fig. 3.6)

$$\tan \alpha = \frac{U(\theta)|\hat{n}| |P_p|}{T(\theta)|P_p|} = \frac{U}{T}(\theta). \qquad (3.79)$$

In general, scattering changes not only the direction but also the magnitude

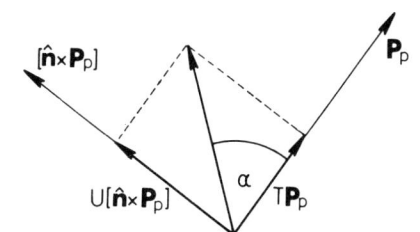

Fig. 3.6. Angle of rotation of the polarization component P_p

of the polarization vector. For example, when $P = 0$, it follows from (3.77) that $P' = S(\theta)\hat{n}$, as already shown in Subsection 3.3.2. Only with $|P| = 1$, that is with total initial polarization, does the degree of polarization of the electrons scattered into a certain direction remain unchanged. This can be seen from formula (3.77) if one uses the relation $S^2 + T^2 + U^2 = 1$, which follows immediately from the definition (3.76):

$$|P'|^2 = \frac{(P_n + S)^2 + T^2 P_p^2 + U^2 P_p^2}{(1 + P_n S)^2},$$

with $|P|^2 = 1$ or $P_p^2 = 1 - P_n^2$ one has

$$|P'|^2 = \frac{P_n^2 + 2P_n S + S^2 + T^2 + U^2 - (T^2 + U^2)P_n^2}{(1 + P_n S)^2}$$

$$= \frac{P_n^2 + 2P_n S + 1 + (S^2 - 1)P_n^2}{(1 + P_n S)^2} = 1.$$

Thus, from $|P|^2 = 1$ it follows that $|P'|^2 = 1$.

In one special case, only the magnitude and not the direction of P is altered by the scattering; this is when $P_p = 0$, or the polarization is perpendicular to the scattering plane. Then from (3.77) it follows that

$$P' = \frac{\{P_n + S(\theta)\}\hat{n}}{1 + P_n S(\theta)}, \tag{3.80}$$

which shows that the polarization retains its direction perpendicular to the scattering plane.

Using the relations given in this section, the quantities S, T, U and thus the scattering amplitudes f and g can be determined experimentally, as will be discussed in Section 3.6. Since $S^2 + T^2 + U^2 = 1$, by measuring two of these parameters one can determine the third, except for the sign. Measurement of these quantities therefore yields only two results which are independent of each other. A third independent result is obtained by measuring the scattering cross section for an unpolarized beam, given by (3.65). Then from (3.65) and (3.76), three of the four parameters of the complex quantities $f = |f| \exp(i\gamma_1)$ and $g = |g| \exp(i\gamma_2)$ can be determined. One can, for example, from $I = |f|^2 + |g|^2$ and $T = (|f|^2 - |g|^2)/(|f|^2 + |g|^2)$ determine $|f|$ and $|g|$. Then from (3.76) the measurement of S and U yields $-2|f||g|\sin(\gamma_1 - \gamma_2)$ and $2|f||g|\cos(\gamma_1 - \gamma_2)$, so that one obtains the difference in the phases of the scattering amplitudes. The complete determination of both phases would contradict the principles of quantum mechanics as it is impossible to determine the absolute phases of wave functions. (For example, the measurement of the cross section $\sigma(\theta) = |f(\theta)|^2$ by the scattering of a particle with zero spin also yields only $|f|$ and not the phase of the scattering amplitude).

Problem 3.4: Calculate P' from (3.74).

Solution: The denominator has already been evaluated in Subsection 3.3.1. The numerator is

$$\tfrac{1}{2} \operatorname{tr} \begin{pmatrix} f & -ge^{-i\phi} \\ ge^{i\phi} & f \end{pmatrix} \begin{pmatrix} 1 + P_z & P_x - iP_y \\ P_x + iP_y & 1 - P_z \end{pmatrix} \begin{pmatrix} f^* & g^*e^{-i\phi} \\ -g^*e^{i\phi} & f^* \end{pmatrix} \begin{pmatrix} \hat{e}_z & \hat{e}_x - i\hat{e}_y \\ \hat{e}_x + i\hat{e}_y & -\hat{e}_z \end{pmatrix}$$

$$= \tfrac{1}{2} \begin{pmatrix} a_{11} & a_{12} \\ a_{21} & a_{22} \end{pmatrix} \begin{pmatrix} \hat{e}_z & \hat{e}_x - i\hat{e}_y \\ \hat{e}_x + i\hat{e}_y & -\hat{e}_z \end{pmatrix}$$

where

$a_{11} = (1 + P_z)|f|^2 - (P_x - iP_y)fg^*e^{i\phi} - (P_x + iP_y)f^*ge^{-i\phi} + (1 - P_z)|g|^2$

$a_{12} = (1 + P_z)fg^*e^{-i\phi} + (P_x - iP_y)|f|^2 - (1 - P_z)f^*ge^{-i\phi} - (P_x + iP_y)|g|^2e^{-2i\phi}$

$a_{21} = (1 + P_z)f^*ge^{i\phi} - (P_x - iP_y)|g|^2e^{2i\phi} + (P_x + iP_y)|f|^2 - (1 - P_z)fg^*e^{i\phi}$

$a_{22} = (1 + P_z)|g|^2 + (P_x - iP_y)f^*ge^{i\phi} + (P_x + iP_y)fg^*e^{-i\phi} + (1 - P_z)|f|^2.$

Hence $\tfrac{1}{2} \operatorname{tr} S(1 + \mathbf{P} \cdot \boldsymbol{\sigma}) S^\dagger \boldsymbol{\sigma}$ equals

$\tfrac{1}{2} \{ [2P_z(|f|^2 - |g|^2) - (P_x - iP_y)e^{i\phi}(fg^* + f^*g) - (P_x + iP_y)e^{-i\phi}(fg^* + f^*g)]\hat{e}_z$

$\quad + [(1 + P_z)fg^*e^{-i\phi} + (P_x - iP_y)(|f|^2 - |g|^2e^{2i\phi}) - (1 - P_z)f^*ge^{-i\phi}$

$\quad - (P_x + iP_y)(|g|^2e^{-2i\phi} - |f|^2) + (1 - P_z)f^*ge^{i\phi} - (1 - P_z)fg^*e^{i\phi}]\hat{e}_x$

$\quad + [(1 + P_z)fg^*e^{-i\phi} + (P_x - iP_y)(|f|^2 + |g|^2e^{2i\phi}) - (1 - P_z)f^*ge^{-i\phi}$

$\quad - (P_x + iP_y)(|g|^2e^{-2i\phi} + |f|^2) - (1 + P_z)f^*ge^{i\phi} + (1 - P_z)fg^*e^{i\phi}]i\hat{e}_y \}$

$= \tfrac{1}{2} \{ [(fg^* - f^*g)(e^{-i\phi} - e^{i\phi}) + P_z(fg^* + f^*g)(e^{-i\phi} + e^{i\phi})$

$\quad + P_x(2|f|^2 - |g|^2(e^{2i\phi} + e^{-2i\phi})) + iP_y|g|^2(e^{2i\phi} - e^{-2i\phi})]\hat{e}_x$

$\quad + [(fg^* - f^*g)(e^{i\phi} + e^{-i\phi}) - P_z(fg^* + f^*g)(e^{i\phi} - e^{-i\phi})$

$\quad + P_x|g|^2(e^{2i\phi} - e^{-2i\phi}) - iP_y(2|f|^2 + |g|^2(e^{2i\phi} + e^{-2i\phi}))]i\hat{e}_y$

$\quad + [2P_z(|f|^2 - |g|^2) - (fg^* + f^*g)(P_x(e^{i\phi} + e^{-i\phi}) - iP_y(e^{i\phi} - e^{-i\phi}))]\hat{e}_z \}$

$= \{ -i(fg^* - f^*g)\sin\phi + P_x(|f|^2 - |g|^2\cos 2\phi) - P_y|g|^2 \sin 2\phi$

$\quad + P_z(fg^* + f^*g)\cos\phi \}\hat{e}_x$

$\quad + \{ i(fg^* - f^*g)\cos\phi - P_x|g|^2\sin 2\phi + P_y(|f|^2 + |g|^2\cos 2\phi) \}$

$$+ P_z(fg^* + f^*g) \sin \phi \} \hat{e}_y$$
$$+ \{-(fg^* + f^*g)(P_x \cos \phi + P_y \sin \phi) + P_z(|f|^2 - |g|^2)\} \hat{e}_z.$$

After rearranging and using $\cos 2\phi = 1 - 2 \sin^2 \phi$, i.e.,

$$P_x(|f|^2 - |g|^2 \cos 2\phi) = P_x(|g|^2 \sin^2 \phi + |g|^2 \sin^2 \phi + |f|^2 - |g|^2),$$
$$P_y(|f|^2 + |g|^2 \cos 2\phi) = P_y(|f|^2 - |g|^2 + |g|^2 \cos^2 \phi + |g|^2 \cos^2 \phi)$$

and $\sin 2\phi = 2 \sin \phi \cos \phi$, we get

$$\tfrac{1}{2} \operatorname{tr} S(1 + \boldsymbol{P} \cdot \boldsymbol{\sigma}) S^\dagger \boldsymbol{\sigma} = (|f|^2 + |g|^2)(-P_x \sin \phi + P_y \cos \phi)(-\sin \phi \hat{e}_x + \cos \phi \hat{e}_y)$$
$$+ i(fg^* - f^*g)(-\sin \phi \hat{e}_x + \cos \phi \hat{e}_y) + (|f|^2 - |g|^2)[(P_x \hat{e}_x + P_y \hat{e}_y + P_z \hat{e}_z)$$
$$- (-P_x \sin \phi + P_y \cos \phi)(-\sin \phi \hat{e}_x + \cos \phi \hat{e}_y)]$$
$$+ (fg^* + f^*g)[P_z \cos \phi \hat{e}_x + P_z \sin \phi \hat{e}_y - (P_y \sin \phi + P_x \cos \phi) \hat{e}_z].$$

With the definitions (3.76) and the relations (3.63) and (3.70) it follows that

$$\boldsymbol{P}' = \frac{[\boldsymbol{P} \cdot \hat{\boldsymbol{n}} + S(\theta)] \hat{\boldsymbol{n}} + T(\theta)[\boldsymbol{P} - (\boldsymbol{P} \cdot \hat{\boldsymbol{n}}) \hat{\boldsymbol{n}}] + U(\theta)[\hat{\boldsymbol{n}} \times \boldsymbol{P}]}{1 + \boldsymbol{P} \cdot \hat{\boldsymbol{n}} S(\theta)}.$$

3.3.4 Double Scattering Experiments

Double scattering experiments are important for determining the Sherman function $S(\theta)$. An unpolarized beam is first scattered by the scattering center 1 (see Fig. 3.7). The electrons scattered through the angles θ_1, $\phi_1 = 0$ [8] are polarized by the scattering process and undergo a second scattering (scattering angles θ_2, $\phi_2 = \phi$, where $\phi = 0$ if the two scattering planes coincide).

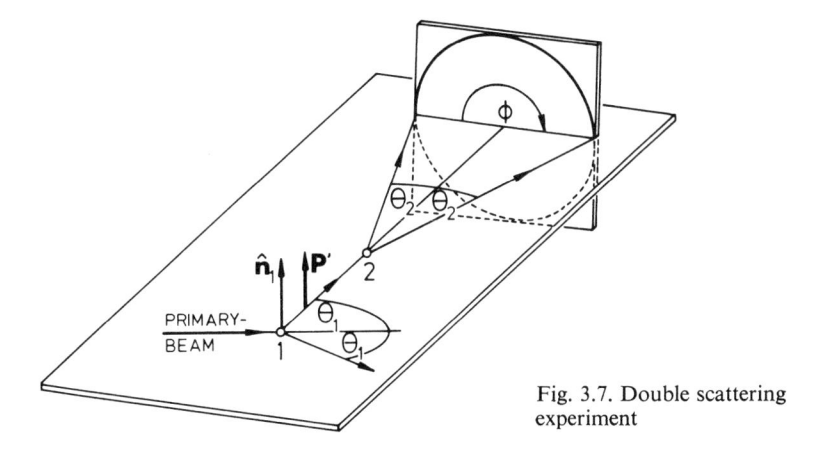

Fig. 3.7. Double scattering experiment

---

[8] As the incident beam is unpolarized, scattering into all azimuthal angles is equally probable, so that we can assign the azimuthal angle $\phi_1 = 0$ to an arbitrary scattering direction.

3. Polarization Effects in Electron Scattering from Unpolarized Targets

From (3.62) the cross section for the scattering by the second target is

$$\sigma_2(\theta_2, \phi_2) = \frac{\operatorname{tr} \rho''}{\operatorname{tr} \rho'},$$

where ρ' and ρ'' are the respective density matrices of the single and double scattered states. One has from (3.71) that $\rho' = \frac{1}{2} SS\dagger \operatorname{tr} \rho$, since the incident beam is unpolarized, and from (3.61) that

$$\rho'' = S'\rho'S'\dagger = \tfrac{1}{2} S'SS\dagger S'\dagger \operatorname{tr} \rho, \tag{3.81}$$

where S and S' are the respective scattering matrices of the first and second scattering processes. Thus we have

$$\sigma_2(\theta_2, \phi_2) = \tfrac{1}{2} \operatorname{tr} S'SS\dagger S'\dagger \frac{\operatorname{tr} \rho}{\operatorname{tr} \rho'}. \tag{3.82}$$

According to Subsection 3.3.1, the cross section for the scattering by the first target where the primary beam is unpolarized is

$$\frac{\operatorname{tr} \rho'}{\operatorname{tr} \rho} = |f(\theta_1)|^2 + |g(\theta_1)|^2 = I(\theta_1).$$

Since $\phi_1 = 0$, one has

$$SS\dagger = \begin{pmatrix} f(\theta_1) & -g(\theta_1) \\ g(\theta_1) & f(\theta_1) \end{pmatrix} \begin{pmatrix} f^*(\theta_1) & g^*(\theta_1) \\ -g^*(\theta_1) & f^*(\theta_1) \end{pmatrix} = I(\theta_1) \begin{pmatrix} 1 & \dfrac{S(\theta_1)}{i} \\ -\dfrac{S(\theta_1)}{i} & 1 \end{pmatrix}.$$

where $S(\theta_1)$ is the Sherman function as defined previously. Hence with $\phi_2 = \phi$ one has

$$\sigma_2(\theta_2, \phi) = \tfrac{1}{2} \operatorname{tr} \begin{pmatrix} f(\theta_2) & -g(\theta_2)e^{-i\phi} \\ g(\theta_2)e^{i\phi} & f(\theta_2) \end{pmatrix}$$

$$\cdot \begin{pmatrix} f^*(\theta_2) - g^*(\theta_2)\dfrac{S(\theta_1)}{i}e^{i\phi} & g^*(\theta_2)e^{-i\phi} + f^*(\theta_2)\dfrac{S(\theta_1)}{i} \\ -f^*(\theta_2)\dfrac{S(\theta_1)}{i} - g^*(\theta_2)e^{i\phi} & -g^*(\theta_2)\dfrac{S(\theta_1)}{i}e^{-i\phi} + f^*(\theta_2) \end{pmatrix}$$

$$= \tfrac{1}{2}\Big[|f(\theta_2)|^2 - f(\theta_2)g^*(\theta_2)\dfrac{S(\theta_1)}{i}e^{i\phi} + f^*(\theta_2)g(\theta_2)\dfrac{S(\theta_1)}{i}e^{-i\phi}$$

$$+ |g(\theta_2)|^2 + |g(\theta_2)|^2 + f^*(\theta_2)g(\theta_2)\dfrac{S(\theta_1)}{i}e^{i\phi}$$

$$- f(\theta_2)g^*(\theta_2)\dfrac{S(\theta_1)}{i}e^{-i\phi} + |f(\theta_2)|^2\Big]$$

$$= I(\theta_2) + \tfrac{1}{2}S(\theta_1)I(\theta_2)S(\theta_2)(e^{i\phi} + e^{-i\phi})$$
$$= I(\theta_2)[1 + S(\theta_1)S(\theta_2)\cos\phi]. \tag{3.83}$$

If the angle between the first and the second scattering plane is zero, we obtain

$$\sigma_2(\theta_2, \phi = 0°) = I(\theta_2)\{1 + S(\theta_1)S(\theta_2)\}.$$

If $\phi = 180°$, then

$$\sigma_2(\theta_2, \phi = 180°) = I(\theta_2)\{1 - S(\theta_1)S(\theta_2)\}.$$

Hence observation of the left-right asymmetry of the intensity in a double scattering experiment yields

$$\frac{\sigma_2(\phi = 0°) - \sigma_2(\phi = 180°)}{\sigma_2(\phi = 0°) + \sigma_2(\phi = 180°)} = S(\theta_1)S(\theta_2). \tag{3.84}$$

If the targets are the same and the scattering angles are equal in the first and second scattering processes ($\theta_1 = \theta_2 = \theta$), one can measure $S^2(\theta)$ for the target in question.

By using the results derived in Subsections 3.3.1 and 3.3.2, the double scattering experiment could have been understood without making further calculations: According to (3.73), the first scattering process produces the polarization $\boldsymbol{P}' = S(\theta_1)\hat{\boldsymbol{n}}_1$. Hence from (3.70) the cross section for the second scattering is

$$\sigma_2 = I(\theta_2)[1 + S(\theta_2)\boldsymbol{P}'\cdot\hat{\boldsymbol{n}}_2] = I(\theta_2)[1 + S(\theta_1)S(\theta_2)\hat{\boldsymbol{n}}_1\cdot\hat{\boldsymbol{n}}_2],$$

where $\hat{\boldsymbol{n}}_1\cdot\hat{\boldsymbol{n}}_2 = \cos\phi$, if ϕ is the angle between the normals of the scattering planes (= angle between the scattering planes).

3.4 Simple Physical Description of the Polarization Phenomena

> The main object of physical science is not the provision of pictures, but is the formulation of laws governing phenomena and the application of these laws to the discovery of new phenomena. If a picture exists, so much the better. ...
>
> P. A. M. DIRAC
> (*The Principles of Quantum Mechanics*, Chapter I,4)

With the use of simple pictures and basic principles, the results which have been mathematically derived in the previous sections will be illustrated. A qualitative explanation will be given for the change in the direction and magnitude of the polarization vector, the asymmetry in the scattering of a polarized electron beam, the fact that the polarization arising from the scattering of an unpolarized beam is perpendicular to the scattering plane, and the fact that the degree of polarization in this case can be described by the same function (Sherman function) which describes the scattering asymmetry of a polarized beam.

3.4.1 Illustration of the Rotation of the Polarization Vector

We shall now illustrate by simple models the results obtained in the last section. First—why does the polarization vector retain its direction only if it is perpendicular to the scattering plane?

The polarization effects in scattering are caused by spin-orbit coupling, in other words, by the magnetic field which the electrons experience in their rest frame (cf. Sect. 3.1): The charged scattering center moves in the rest frame of the electrons; the current that is represented by this moving charge produces a magnetic field $\boldsymbol{B} = \boldsymbol{E} \times \boldsymbol{v}/c$ which acts upon the magnetic moments of the electrons. As $\boldsymbol{E}$ and $\boldsymbol{v}$ lie in the scattering plane, $\boldsymbol{B}$ is perpendicular to this plane (see Fig. 3.8). If the polarization $\boldsymbol{P}$ does not lie parallel or antiparallel to $\boldsymbol{B}$, the magnetic moment which is connected with $\boldsymbol{P}$ experiences a torque that induces $\boldsymbol{P}$ to change its direction and to precess. Only if the polarization is parallel or antiparallel to $\boldsymbol{B}$ does it retain its direction.

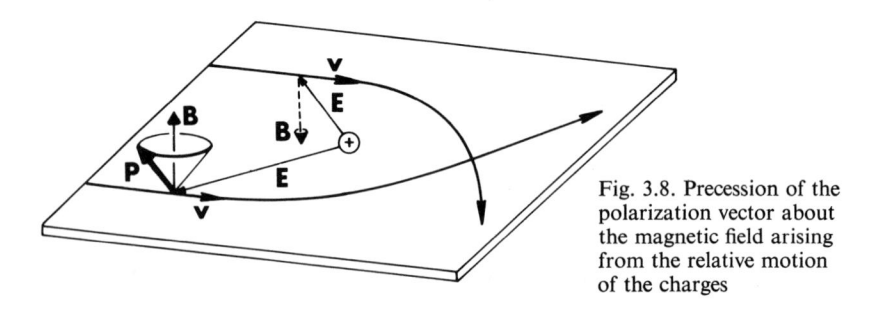

Fig. 3.8. Precession of the polarization vector about the magnetic field arising from the relative motion of the charges

3.4.2 Illustration of the Change in the Magnitude of the Polarization Vector

The picture just used does not answer the question of how it is possible for the magnitude of $\boldsymbol{P}$ to change during the scattering. We first explain this for the conspicuous case in which the degree of polarization changes from zero to a finite value (scattering of an unpolarized beam).

For an unpolarized beam, if we take 300 eV incident energy and scattering by Hg as an example, the scattering cross section has the shape shown

3.4 Simple Physical Description of the Polarization Phenomena

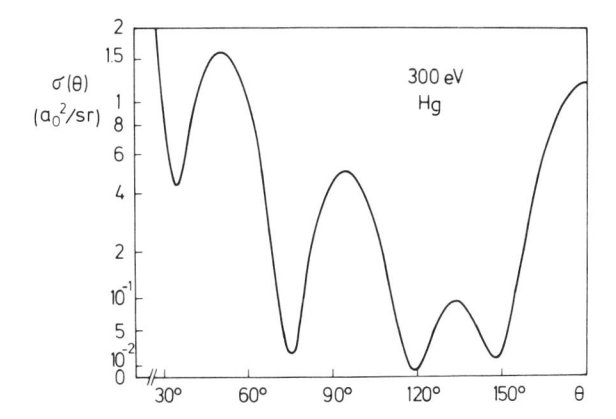

Fig. 3.9. Differential cross section for elastic scattering of an unpolarized electron beam (a_0 = Bohr radius). Ordinate pseudologarithmic according to $\log[1 + 10\sigma(\theta)/(a_0^2/\text{sr})]$

in Fig. 3.9. The typical interference structure of the cross section arises because the electron wavelength is of the same order of magnitude as the atomic radius ($\lambda = 1$ Å for 150 eV).

It follows from Chapter 2 that one can consider the unpolarized beam as a mixture of two equal fractions with opposing spin directions. It is expedient to choose the arbitrary spin directions of these two constituent beams to be perpendicular to the scattering plane because they then remain unchanged in the scattering process.

The cross sections of the two beams with opposite polarizations differ slightly from each other. This is because the scattering potential essentially consists of the electrostatic and the spin-orbit potential: $V = V_0 + V_{ls}$. Since V_{ls} contains the scalar product $\boldsymbol{l} \cdot \boldsymbol{s}$, it has different signs for electrons of the same orbit but different spin directions. As Fig. 3.10 shows, the resulting scattering potential will therefore be higher or lower for spin-up electrons (e↑) than for spin-down electrons (e↓), depending on which side of the atom they pass. If we consider, for example, electrons which pass by the atom on the right, i.e., that are scattered to the left, it can be seen from Fig. 3.10 that the effective radius R of the atom for scattering (defined as the radius where the potential has dropped to a certain value) will be smaller for e↑ than for e↓. Since the positions of the extrema in interference patterns like those shown in Fig. 3.9 are determined by λ/R, their abscissas depend on the effective atomic radius, so that different cross-section curves result for e↑ and e↓.[9] Since a change in the scattering potential affects also

---

[9] Needless to say, quantitative results cannot be derived from these qualitative arguments.

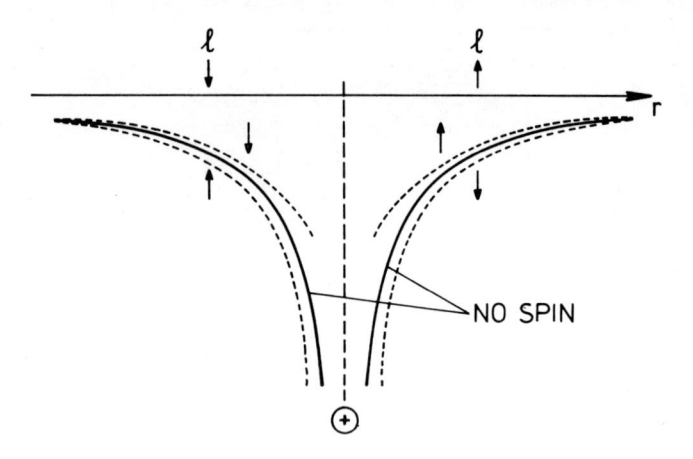

Fig. 3.10. Potential curves with (---) and without (—) spin-orbit coupling for electrons with spins up ↑ and down ↓

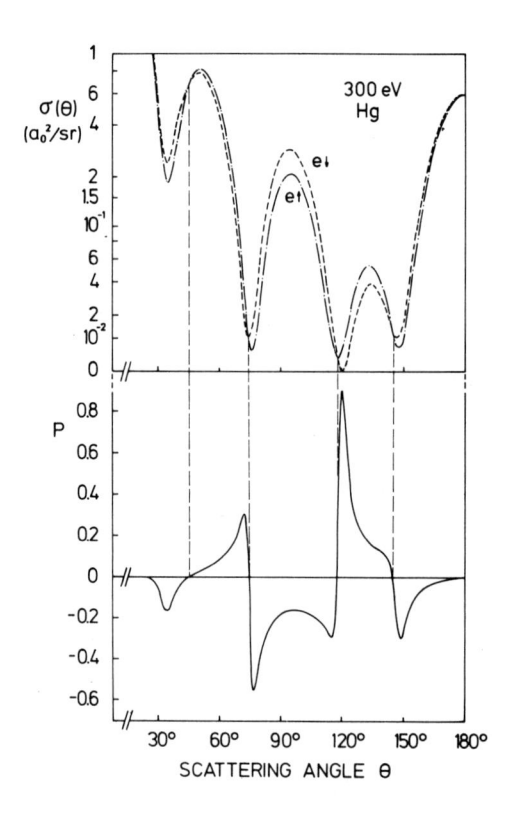

Fig. 3.11. Construction of the polarization from the cross sections for e↑ and e↓. Ordinate of cross sections pseudologarithmic according to $\log[1 + 50\sigma(\theta)/(a_0^2/\mathrm{sr})]$

the ordinates of the cross sections, one obtains the curves shown in the upper half of Fig. 3.11 for scattering to the left.

The numbers of e↑ and e↓ scattered in a particular direction θ are therefore usually different from each other; in other words, the scattered beam is polarized. Since the number of scattered electrons is proportional to the corresponding cross section, the degree of polarization is

$$P = \frac{N_\uparrow - N_\downarrow}{N_\uparrow + N_\downarrow} = \frac{\sigma_\uparrow - \sigma_\downarrow}{\sigma_\uparrow + \sigma_\downarrow}, \quad (3.85)$$

so that one can construct P directly from the cross-section curves as indicated in Fig. 3.11.

What has been shown here for a mixture of 50% e↑ and 50% e↓ (unpolarized beam) is also valid for every other mixture (partially polarized beam, $P < 1$): Since the cross sections for e↑ and e↓ are different, the proportions of the mixture change with scattering, which means, according to (3.85), that the degree of polarization changes.

From the construction just described it follows that a particularly high polarization arises at those angles where one of the two cross sections has a deep minimum so that its value is very small compared to that of the other cross section at the same angle. Electrons of a single spin direction then predominate in the scattered beam, so that one approaches the ideal case of a totally polarized electron beam. Due to the fact that the spin-orbit interaction is relatively weak, the mutual shift of $\sigma_\uparrow$ and $\sigma_\downarrow$ is not significant (see Fig. 3.11), so that the complete differential cross section $\sigma(\theta) = \sigma_\uparrow(\theta) + \sigma_\downarrow(\theta)$ likewise has its minima near the extreme values of P. This would be different if the spin-orbit coupling were so large that $\sigma_\uparrow$ and $\sigma_\downarrow$ were strongly shifted with respect to each other. Such cases occur in nucleon scattering. For electron scattering we can, however, note that high degrees of polarization occur only near cross-section minima.

3.4.3 Illustration of the Asymmetry in the Scattering of a Polarized Beam

The left-right asymmetry which is observed in the scattering of a polarized electron beam can immediately be interpreted with the help of Fig. 3.10. We assume without loss of generality that the beam consists only of e↑. Then the electrons that pass the atom on the left, i.e., that are scattered to the right, experience a stronger potential than those scattered to the left. Different scattering potentials cause different scattering intensities, so that one observes a scattering asymmetry.

A quantitative example can be taken from the upper part of Fig. 3.11 which shows the cross sections for scattering of e↑ and e↓ to the left. Remembering the discussion in Subsection 3.4.2, we see that this graph also represents the case of scattering to the right, if one interchanges the labels ↑ and ↓. The two cross sections in Fig. 3.11 can therefore also be taken to represent the scattering of e↑ to the left and to the right, respectively. This shows that there is a left-right asymmetry of the scattering intensity—except for those angles where the cross sections happen to intersect.

3.4.4 Transversality of the Polarization as a Consequence of Parity Conservation. Counterexample: Longitudinal Polarization in β Decay

The polarization resulting from the scattering of an unpolarized electron beam is perpendicular to the scattering plane, as proved in Subsection 3.3.2. This can be seen directly from simple symmetry considerations. According to all physical experience, parity is conserved for the electromagnetic interaction which governs electron scattering. In other words, the result of an electron-scattering experiment must not depend on whether it is described in a right- or left-handed (e.g., reflected) coordinate system. This means that the mirror image of such an experiment must represent a possible course of the experiment, too.

From this principle it follows that the polarization P of the scattered beam cannot have a longitudinal component $P \cdot k_2$. Otherwise this component would define a certain screw sense (helicity) in the laboratory system, for example, a right-handed screw, $P \cdot k_2 > 0$. With the reflection of the experiment (see Fig. 3.12) one would obtain a left-handed screw, $P \cdot k_2 < 0$. This is because the sense of rotation connected with the polarization is retained with the reflection, whereas the momentum direction k_2 is inverted. Hence, with the same initial state, the mirror image yields a different final state. This can clearly be seen if one thinks of the image as

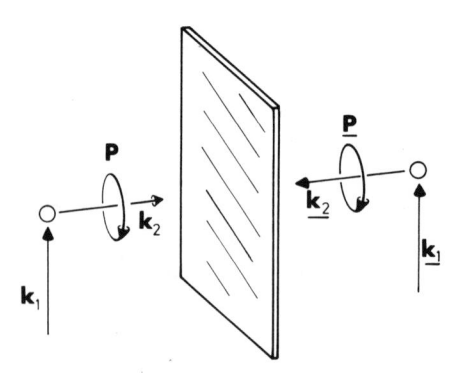

Fig. 3.12. Reflection of a scattering process with unpolarized initial state. To avoid confusion, the axial vector P is depicted by its direction of rotation

3.4 Simple Physical Description of the Polarization Phenomena   57

being rotated through 180° about k_1 and imposed on the original experiment. Then the momenta coincide, but there is a difference in the sense of rotation. As we must expect a well-defined final state when we have a well-defined initial state, this course of the experiment contradicts reality.

From the same argument it follows that a component $P \cdot k_1$ in the incident direction cannot exist either. This means that P must be perpendicular to k_1 and k_2, i.e., perpendicular to the scattering plane.

The mirror inversion being considered $(x, y, z) \to (x, -y, z)$ is not identical to the parity inversion $(x, y, z) \to (-x, -y, -z)$. It can, however, be transformed to this by a rotation through 180° (see Fig. 3.13). As the screw sense is not changed by a rotation, use of the simple mirror inversion is justified for our considerations.

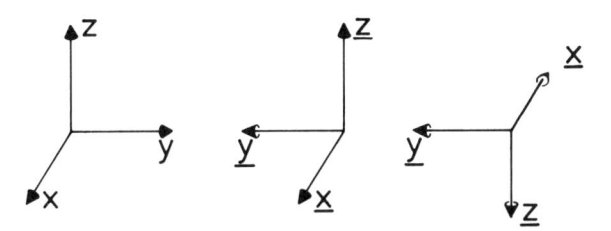

Fig. 3.13. Mirror inversion followed by a rotation through 180° yielding the parity inversion

To prevent misunderstanding, it must be emphasized that disappearance of the longitudinal polarization component can no longer be inferred from parity conservation, if one has a non-zero initial polarization. We explain this by an example. Let the initial polarization be transverse but parallel to the scattering plane (see Fig. 3.14). Let us further assume that

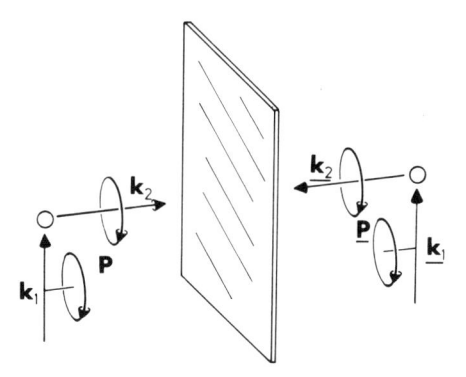

Fig. 3.14. Reflection of a scattering process with polarized initial state

the spin-flip amplitude is small so that the polarization vector is virtually not changed by the scattering, as is actually often the case (cf. Sect. 3.5). Then P approximately retains its direction in space. In the example shown, the right-handed screw formed by P and k_2 in the laboratory is transformed into a left-handed screw in the mirror image. In this case, however, the mirror image also represents an experiment that is possible in nature; we could have performed it by setting the detector in the laboratory to the left instead of to the right of the scatterer, and would have obtained the same result, since there is as much scattering to the left as to the right. We can also think of the mirror image as being rotated through 180° about k_1 and imposed on the left-hand side. This again produces coincident momenta and a different sense of rotation. In this case, however, it is not only the final state that has been changed by the reflection but also the initial state; therefore the image represents a possible course of the experiment.

With interactions that violate parity conservation, arguments of the kind just used cannot be applied. Parity violation can then even be the reason for the spin polarization of the particles concerned. A famous example of this is β decay.

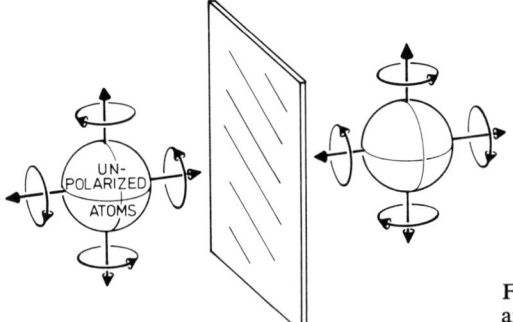

Fig. 3.15. Radioactive β source and its mirror image

A longitudinal polarization of the electrons emitted by unpolarized nuclei, as actually occurs in β decay, is incompatible with parity conservation. Fig. 3.15 shows that all left-handed screws are transformed to right-handed screws by reflection, so that the final states in the mirror image and in the laboratory differ from each other although the initial states are the same. Accordingly, the result of the experiment is not invariant under spatial inversion; in other words, the law of conservation of parity is violated.

The currents of polarized electrons which one can obtain from radioactive sources are very small. Electron polarization in β decay is, however, of fundamental importance as it led to the discovery of parity violation in weak interactions. The theoretical and experimental aspects of β decay are treated in detail in modern textbooks on nuclear physics and in review articles so that we will not consider them further [3.5].

Problem 3.5: It was shown in Subsection 3.3.1 that the polarization components which lie in the scattering plane do not contribute to the scattering asymmetry. Derive this result from symmetry considerations.

Solution: We resolve the polarization component P_p which lies parallel to the scattering plane (see Fig. 3.5) into a component $P_\parallel$ in the incident direction and a component $P_\perp$ perpendicular to it (see Fig. 3.16). The former, due to the rotational symmetry, cannot give rise to an asymmetrical intensity distribution (cf. Example 3.2). Let us assume that $P_\perp$ causes different intensities in the directions k_2 and k_2' which are symmetrical to the incident direction. The reflection through the incident direction (mirror perpendicular to scattering plane) would then invert the scattering asymmetry but leave $P_\perp$ (sense of rotation!) unchanged. Hence the final state represented by the mirror image would differ from the final state in the laboratory although the initial states are the same. Accordingly, the result of the experiment would depend on whether it is described in the original or the reflected coordinate system.

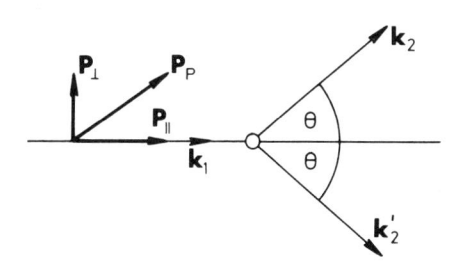

Fig. 3.16. Resolution of the polarization component P_p. (Scattering plane = plane of diagram)

3.4.5 Equality of Polarizing and Analyzing Power

We now give the connection between the following two facts which were proved in Section 3.3:

a) Scattering as a polarizer: A beam that is originally unpolarized obtains the polarization $P = S(\theta)\hat{n}$ from scattering, i.e., the polarization is determined by the Sherman function.

b) Scattering as an analyzer: The left-right asymmetry observed with the scattering of a beam that is polarized perpendicular to the scattering plane is also determined by $S(\theta)$. For a totally polarized beam one has

$$\frac{N_l - N_r}{N_l + N_r} = \frac{1 + S(\theta) - (1 - S(\theta))}{1 + S(\theta) + 1 - S(\theta)} = S(\theta) \qquad (3.86)$$

(N_l, N_r = number of electrons scattered to left and right, respectively).

Polarizer and analyzer are thus characterized here by one and the same function.[10] It can easily be seen that this must be true because the first fact follows immediately from the second.

Once again we consider the incident unpolarized beam to be a mixture of equal numbers of electrons polarized in opposite directions. One half are totally polarized in the direction ↑ perpendicular to the scattering plane; the other half are totally polarized in the opposite direction (see Fig. 3.17). From (3.70) it follows for the e↑ beam that the scattering intensity to the left is proportional to $1 + S(\boldsymbol{P}\cdot\hat{\boldsymbol{n}}) = 1 + S$, whereas the intensity to the right is proportional to $1 - S$ due to the opposite vector $\hat{\boldsymbol{n}}$. For the e↓ beam the corresponding values are $1 - S$ and $1 + S$. In both cases the polarization vectors remain unchanged as they are perpendicular to the scattering plane and have the magnitude 1 (cf. Subsects. 3.3.3 and 3.4.1).

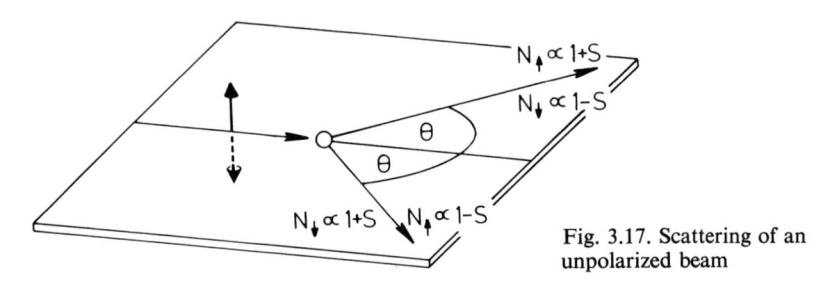

Fig. 3.17. Scattering of an unpolarized beam

The fact that there are different numbers of e↑ and e↓ in the scattered beam means that this beam is polarized. Figure 3.17 illustrates that for scattering to the left the degree of polarization is

$$P = \frac{N_\uparrow - N_\downarrow}{N_\uparrow + N_\downarrow} = \frac{1 + S - (1 - S)}{1 + S + 1 - S} = S,$$

whereas for scattering to the right one has $P = -S$ (as long as ↑ and ↓ always refer to the same reference direction). The polarization after the scattering of an unpolarized beam is thus seen to be a direct consequence of the scattering asymmetry of a polarized beam. Both quantities are described by the same function $S(\theta)$.

Problem 3.6: In the preceding considerations, the incident unpolarized beam was separated into two beams which were polarized perpendicular to the scattering plane. Show that the result does not depend on this particular separation, i.e., that the separation can be made in any arbitrary direction.

---

[10] That this is not so with all scattering processes is shown in Problem 4.1.

Solution: We choose an arbitrary separation of the incident beam. Let one half have the polarization $\boldsymbol{P} = (\boldsymbol{P}_n, \boldsymbol{P}_p)$ and the other half $-\boldsymbol{P} = (-\boldsymbol{P}_n, -\boldsymbol{P}_p)$. The number N_+ of electrons from the part of the beam with polarization $\boldsymbol{P}$ which are scattered at the angle θ is proportional to $1 + S(\theta)\boldsymbol{P}\cdot\hat{\boldsymbol{n}} = 1 + S(\theta)P_n$. The polarization of these electrons after scattering is, according to (3.77),

$$\boldsymbol{P}'_+ = \frac{[P_n + S(\theta)]\hat{\boldsymbol{n}} + T(\theta)\boldsymbol{P}_p + U(\theta)[\hat{\boldsymbol{n}} \times \boldsymbol{P}_p]}{1 + P_n S(\theta)}.$$

The number N_- of electrons from the part of the beam with polarization $-\boldsymbol{P}$ which are scattered at the same angle is proportional to $1 - S(\theta)\boldsymbol{P}\cdot\hat{\boldsymbol{n}} = 1 - S(\theta)P_n$. The polarization vector of these electrons after scattering is

$$\boldsymbol{P}'_- = \frac{[-P_n + S(\theta)]\hat{\boldsymbol{n}} - T(\theta)\boldsymbol{P}_p - U(\theta)[\hat{\boldsymbol{n}} \times \boldsymbol{P}_p]}{1 - P_n S(\theta)}.$$

The resultant polarization $\boldsymbol{P}'$ of the scattered beam is according to (2.16)

$$\boldsymbol{P}' = \frac{N_+}{N_+ + N_-}\boldsymbol{P}'_+ + \frac{N_-}{N_+ + N_-}\boldsymbol{P}'_- = \frac{2S(\theta)\hat{\boldsymbol{n}}}{2} = S(\theta)\hat{\boldsymbol{n}}.$$

Again one has a beam polarized perpendicularly to the scattering plane with degree of polarization $S(\theta)$.

3.5 Quantitative Results

Numerical values are given for the quantities characterizing the scattering process. They show that if the atomic number of the target and the scattering angle are not too small, significant polarization effects occur. At energies of less than 100 eV the quantitative results are still incomplete and not very accurate.

Until now we have shown only which basic phenomena arise in Mott scattering and have said nothing about their magnitude. All that one needs to assess them are the complex amplitudes f and g since the scattering process is completely determined by these amplitudes. Apart from depending on the scattering angle θ, they depend also on the energy of the incident electrons and on the scatterer (i.e. essentially on the scattering potential).

3.5.1 Coulomb Field

Even for the simplest case, the scattering in the pure Coulomb field of the atomic nucleus, the calculation of f and g is tedious. The Dirac equation for the Coulomb field can be exactly solved so that the scattering phases η_l and η_{-l-1} in the infinite series for f and g [Eq. (3.51)] can be given exactly. However, closed expressions for the series cannot be obtained. Therefore one finds, apart from approximation formulae (e.g., from McKinley and Feshbach for the scattering cross section) only numerical tables for the quantities of interest [3.6].

3. Polarization Effects in Electron Scattering from Unpolarized Targets

As measurements in the past 10–20 years have shown, these theoretical results for the pure Coulomb field are very reliable. They are only valid, however, when the electron energies and scattering angles are not too small. This can be easily visualized: Slow electrons are strongly deflected even by the relatively weak forces at large distances from the nucleus. At these large distances the screening of the nuclear Coulomb field by the electron cloud is very effective. Also for the scattering of fast electrons at small angles, the relatively weak forces at large distances from the nucleus play the major role. In order to be scattered through large angles, however, fast electrons must come very close to the atomic nucleus, so that in the region of a few hundred keV with scattering angles of more than 45°, the description of the scattering process by the Coulomb field is a good approximation.

As an example, Fig. 3.18 shows the cross section for the scattering of unpolarized 204-keV electrons by mercury, divided by the well-known Rutherford cross section. If the Rutherford formula were exact for electron scattering, one would obtain a horizontal straight line with the ordinate 1 for this scattering process. The actual scattering cross section deviates from this straight line at angles above about 30° (solid curve) because the Rutherford formula neglects the influence of the spin-orbit interaction. The very strong decrease of the cross section at small angles (dashed curve) is due to the screening of the nuclear Coulomb field by the electron cloud of the atom.

Figure 3.19 shows the Sherman function in the region where scattering by the Coulomb potential is a very good approximation. It can be seen

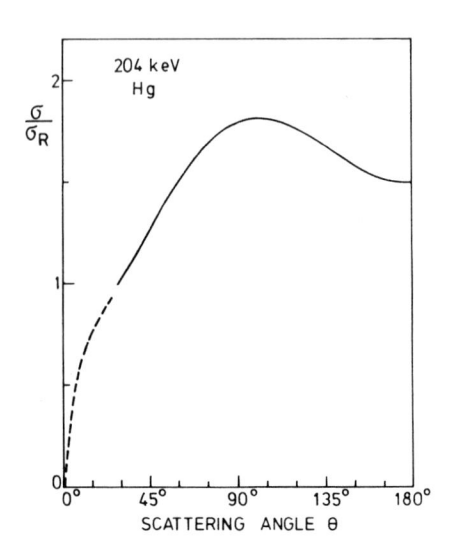

Fig. 3.18. Ratio of the cross section for elastic scattering to the Rutherford cross section for 204-keV electrons scattered by Hg [3.7]

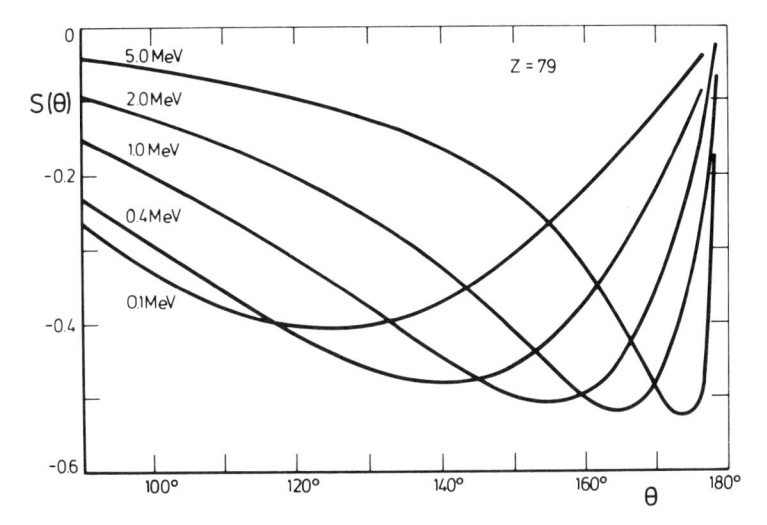

Fig. 3.19. Sherman function $S(\theta)$ for scattering by gold at various energies [3.6]

that with suitable electron energies and large scattering angles, polarizations of 40–50% occur when an unpolarized beam is scattered by gold. The fairly large values of $S(\theta)$ also imply that the left-right asymmetry in scattering of polarized electrons is easily detectable. Such appreciable values occur only with elements having high atomic numbers since only then is the spin-orbit coupling large enough to evoke significant spin effects. With light elements the polarization effects are vanishingly small. A complete survey of theoretical and experimental work in the higher range of energies discussed here was given by ÜBERALL [3.8].

3.5.2 Screened Coulomb Field

The theoretical calculation of the scattering amplitudes f and g for low energies is much more tedious as it can no longer be based on the Coulomb potential. Very reliable results were obtained by using Hartree-type potentials and evaluating f and g by machine computation. We will now discuss these results.

The typical behavior of the cross sections in this energy region was already shown in Fig. 3.9 for 300-eV electrons. The curves are not smooth as in the 100-keV region but have an interference structure as λ is of the same order as the atomic radius. As explained in Subsection 3.4.2, the polarization curves [i.e., $S(\theta)$] are closely connected to the cross-section curves. They therefore also have an oscillating character and are not as

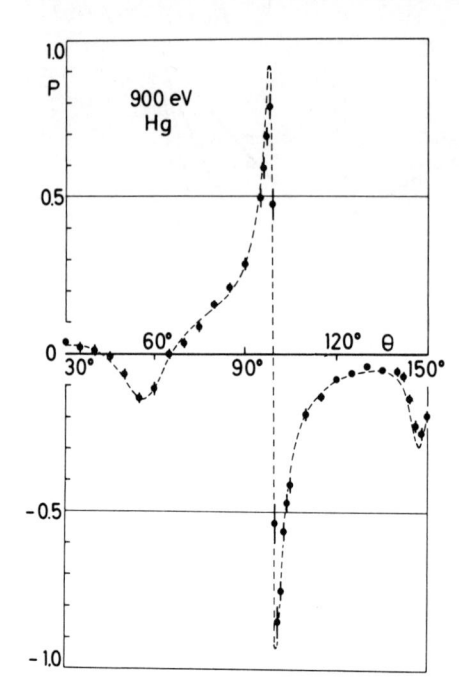

Fig. 3.20. Sherman function for scattering of 900-eV electrons by Hg [3.9]

smooth as in the region of a few hundred keV (Fig. 3.19). As an example, Figure 3.20 shows the Sherman function for the scattering of 900-eV electrons by Hg in the angular range where high values of $S(\theta)$ occur. Small angles can always be disregarded when considering the polarization effects since the Sherman function is then practically zero. Since the cross sections are very large at small angles, this is again in accordance with our rule of Subsection 3.4.2 that high values of the Sherman function occur where the cross section is small and vice versa. It is also to be intuitively expected that there are no high polarization effects at small angles: Electrons that are scattered at small angles have passed through the relatively weak atomic field at fairly large distances from the nucleus. There the spin-orbit interaction which causes the polarization effects is very small, since, according to Section 3.1, it decreases quickly with increasing distance from the nucleus.

A good survey of polarization effects is given by the contours $S(\theta, E) = $ const. Figure 3.21 shows that for certain energy-angle combinations high values ($|S| > 0.8$) of the Sherman function occur. The positions of the extrema of $S(\theta, E)$ have been determined by WALKER [3.10] using a method given by BÜHRING [3.11]. Such high values are of interest when scattering is used as a polarizer or when polarization is analyzed by measuring the

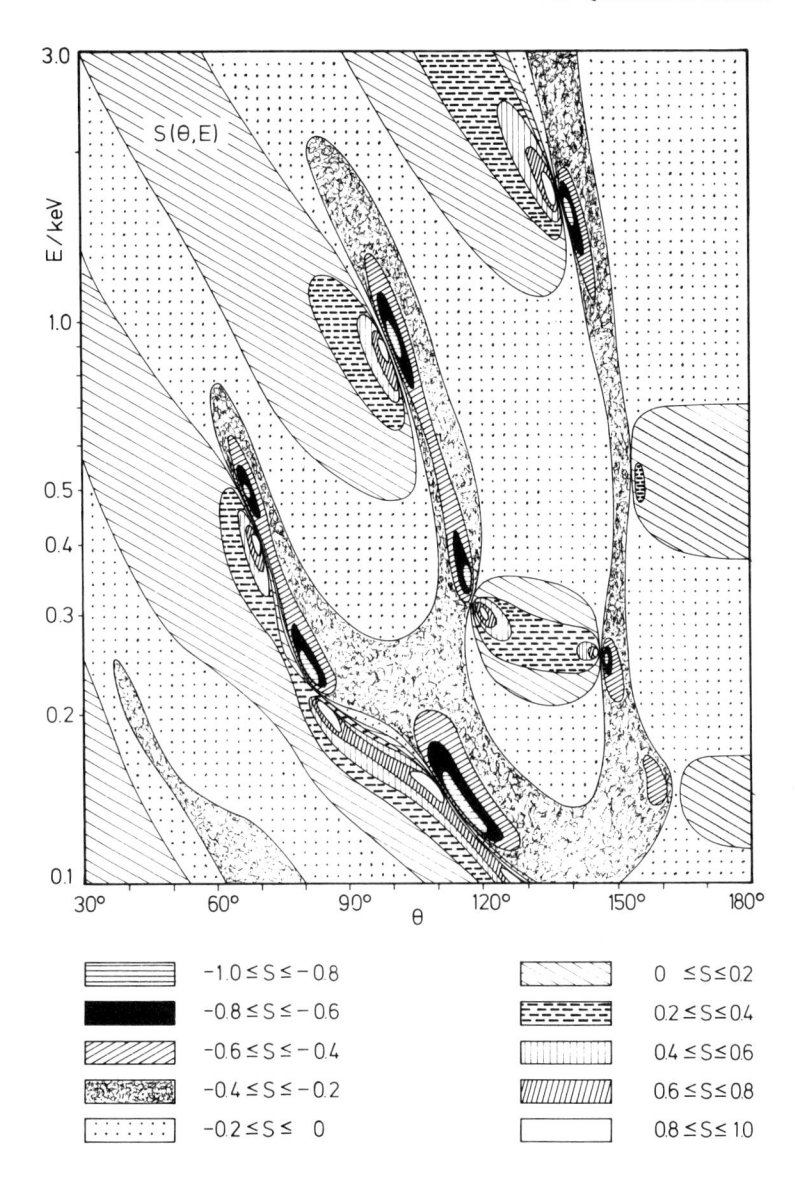

Fig. 3.21. Contours $S(\theta, E)$ = const. for Hg. (Logarithmic energy scale)

66   3. Polarization Effects in Electron Scattering from Unpolarized Targets

left-right asymmetry. One must, however, remember that these "favorable" parameter combinations lie near the cross-section minima. One therefore obtains small scattering intensities and thus little efficiency (cf. Subsect. 3.6.3 and Sect. 7.3). Furthermore, the extrema with values of $|S|$ near 1 are extremely narrow (see Fig. 3.20). Accordingly, $|S| \approx 1$ can be realized experimentally only if one works with very good angular resolution, which again means a reduction of intensity.

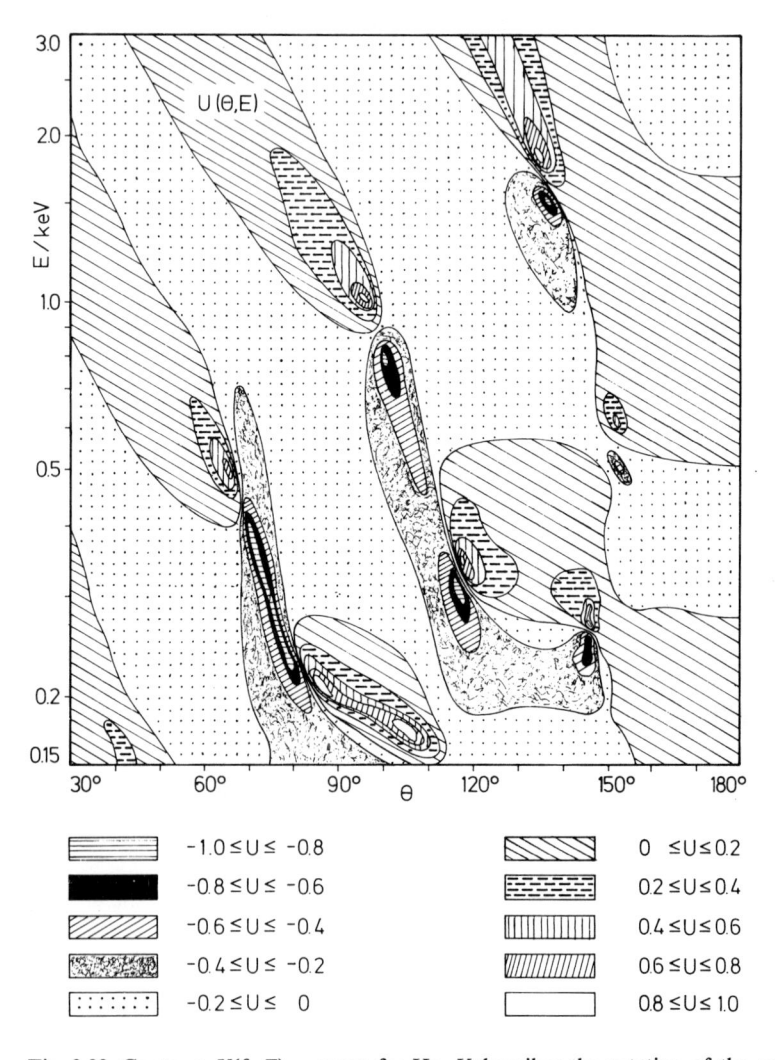

Fig. 3.22. Contours $U(\theta, E)$ = const. for Hg. U describes the rotation of the polarization out of its initial plane [see (3.77)]

In Fig. 3.22 the corresponding contours for $U(\theta, E)$ are shown. The function U describes the rotation of the polarization vector out of its initial plane (cf. Subsect. 3.3.3). The nearer $|U|$ is to 1, the stronger is this rotation of $\boldsymbol{P}$ during the scattering. The figure shows that for certain energy-angle combinations very strong rotation occurs.

Figure 3.23 shows the corresponding contours for the function $T(\theta, E)$, which according to (3.77) describes the reduction of the polarization component $\boldsymbol{P}_p$ parallel to the scattering plane. The more $T(\theta, E)$ deviates from

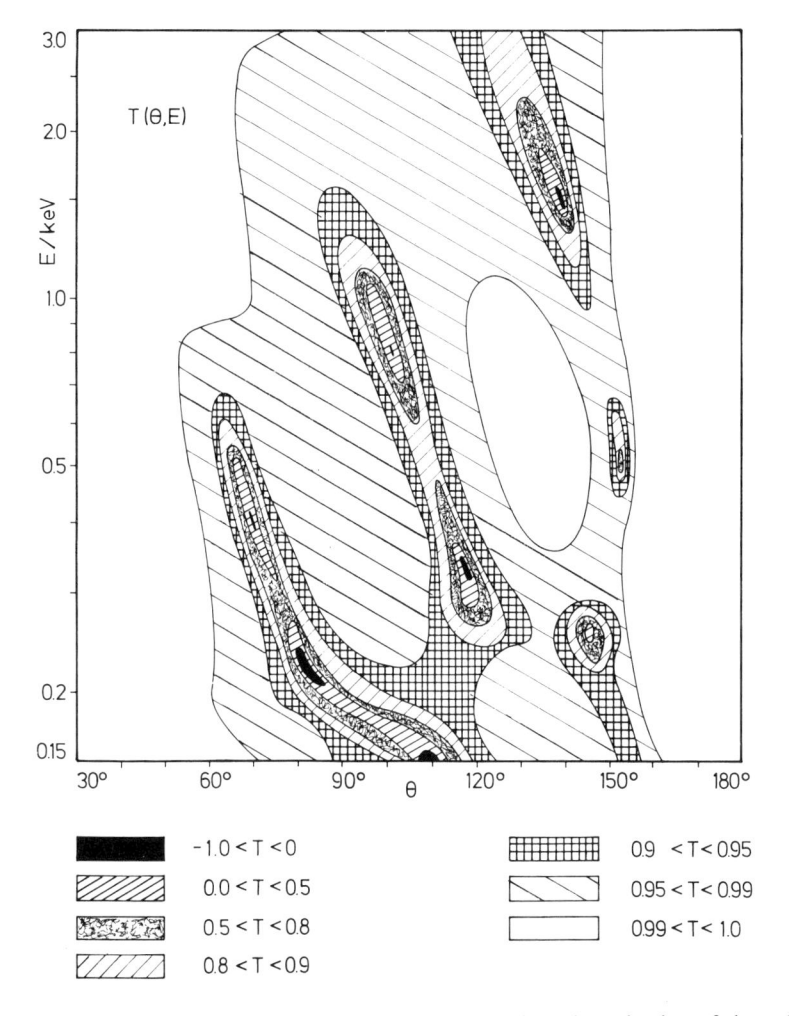

Fig. 3.23. Contours $T(\theta, E) =$ const. for Hg. T describes the reduction of the polarization component $\boldsymbol{P}_p$ due to scattering [see (3.77)]

+1, the stronger is this reduction. Here again, strong changes of the polarization for certain parameter combinations can be seen.

A comparison of theoretical and experimental results shows that the theoretical prediction of the polarization effects is reliable down to the region of about 100 eV. The Sherman function for scattering angles between 30° and 150° was determined experimentally in the energy region shown in Fig. 3.21 so that the errors of these curves are well known. The error limits, even for the extrema, are usually only a few percent, and for the smaller energies shown occasional small angular shifts of up to 2° near the extrema cannot be excluded. With energies smaller than those shown in Fig. 3.21, deviations between theory and experiment continually increase [3.10, 12].

There are no experimental values at all for T and U in the energy region of Figs. 3.22 and 3.23. At the lower energies shown there, deviations occur between the different theories. Since these deviations have not yet been clarified experimentally, these figures have not been drawn for quite such low energies as Fig. 3.21.

The uncertainty of the theoretical values at low energies is due to the fact that with decreasing energy, processes become important which are difficult to describe theoretically. First, a charge distortion (and thus a potential distortion) arises in the atomic electron cloud during the scattering process, because slow electrons stay near the atom long enough to give rise to such a charge-cloud polarization (not to be confused with spin polarization). Second, at low energies exchange processes of the incident electrons with the atomic electrons are important. Further work at energies below 100 eV is necessary.

We have confined our discussion to heavy elements, such as mercury and gold. Polarization effects also exist with lighter elements as targets [3.13], but due to the smaller spin-orbit coupling they are less pronounced.

Due to the introductory character of this book we must restrict ourselves to a few typical examples. A complete survey of the existing results is given in a review article by WALKER [3.10] from the theoretical standpoint, and in a review article by the author [3.14]. A considerable amount of material is also given in a review by ECKSTEIN [3.15]. Tables of the relevant functions have been published by HOLZWARTH and MEISTER [3.16], and by FINK, YATES et al. [3.17].

3.6 Experimental Investigations

Typical examples of experimental setups for double and triple scattering experiments are given, and the main difficulties in carrying out such measurements (small intensity, elimination of unwanted electrons, plural scattering) are discussed. The polarization

3.6.1 Double Scattering Experiments

In polarization experiments one usually does not have just one scattering process as in the measurements of cross sections. According to Subsection 3.3.4, one can for example measure the Sherman function $S(\theta)$ by the following method. An unpolarized beam is scattered through the angle θ' (Fig. 3.24). The scattered beam which has the polarization $P = S(\theta')$ then hits a second target of the same kind. The respective numbers of electrons N_1 and N_r scattered through the same angle θ' to the left and to the right are measured. From relation (3.84) it follows that

$$\frac{N_1 - N_r}{N_1 + N_r} = S^2(\theta')$$

so that one obtains the value of $|S|$ for the angle θ'.

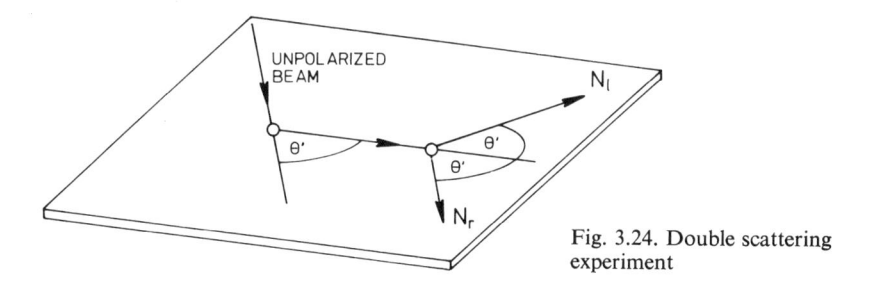

Fig. 3.24. Double scattering experiment

The angular dependence $S(\theta)$ can be found if one varies the first scattering angle and leaves the scattering angle at the second target unchanged. According to (3.84) one then measures

$$\frac{N_1 - N_r}{N_1 + N_r} = S(\theta)S(\theta').$$

As $|S(\theta')|$ is known from the first experiment, this measurement yields $S(\theta)$. Strictly speaking, the sign of $S(\theta)$ is not determined because only the magnitude of $S(\theta')$ is known. By using a source that produces polarized electrons of known spin direction, one can, however, determine the sign of $S(\theta')$ by measuring the sign of the left-right asymmetry which arises in the scattering of these electrons through θ' [see (3.70)].

The main problem in carrying out precise measurements of the Sherman function is to obtain large enough intensities in the detectors for N_l and N_r, since a scattered beam which has a weak intensity has to be scattered again. The polarized electrons which one needs for a measurement of the Sherman function could alternatively be produced by processes other than scattering (cf. Sect. 7.3). These processes, however, do not yield enough intensity to invalidate our remark on the intensity problem. To date there exists no source of polarized electrons whose intensity is even approximately comparable with that of normal electron sources.

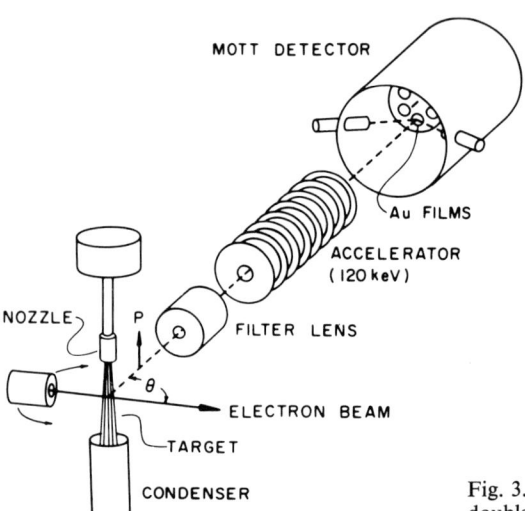

Fig. 3.25. Experimental setup of a double scattering experiment [3.9]

Figure 3.25 shows an example of a practical setup for a double scattering experiment which has been used for a large number of measurements. The electron gun yields a well-collimated beam of unpolarized electrons of fixed energy E_0. The energy E_0 lies between 25 eV and a few keV. The electrons are fired at a mercury-vapor target and the degree of polarization $P(\theta, E_0)$ of the scattered electrons is measured. As we are considering the scattering of an initially unpolarized beam, we have $P(\theta, E_0) = S(\theta, E_0)$. Thus the measurement produces the Sherman function. The scattering angle θ is varied by rotating the gun about the axis of the Hg beam.

The scattered beam passes through a filter lens, which removes inelastically scattered electrons, because in the present example elastic scattering is to be studied. The filter lens not only removes those electrons that excited the Hg atoms into higher energy states, but also removes unwanted electrons that hit the walls of the vacuum chamber (not shown) and were reflected into the direction of observation, having lost some of their energy.

The electrons that leave the filter lens are accelerated to 120 keV and then hit the second target, a gold foil. The electrons scattered through 120° to the right and left are counted. From the left-right asymmetry

$$\frac{N_1 - N_r}{N_1 + N_r} = S(\theta, E_0)S(120°, 120 \text{ keV}) \tag{3.87}$$

one obtains the quantity $S(\theta, E_0) = P(\theta, E_0)$ (i.e., the polarization after the first scattering process) if $S(120°, 120 \text{ keV})$ is known. Accelerating the electrons to 120 keV after the first scattering process has the advantage that at this energy the Sherman function is particularly well known. The theoretical values are very reliable as the approximation of scattering in the pure Coulomb field of the nucleus is very good in this energy region (see Sect. 3.5). Furthermore, the experimental determination of the Sherman function has been carried out particularly accurately at this energy [3.18]. This is partly because extensive Mott-scattering investigations of electron polarization in β decay took place in this energy region.

We emphasize once more the main problems of such experiments because these problems have not always been fully appreciated, the experiments thus having led to erroneous results.

Since the intensity of the electrons that are to be detected is small, the background of unwanted electrons must be carefully suppressed. This background appears because neither the electrons of the primary beam nor scattered electrons that hit the walls of the scattering chamber or of the polarization analyzer are completely absorbed there. They are instead reflected at the walls, and an appreciable portion of them, if not sufficiently suppressed, arrives at the counters and affects the measurements.

Electrons that are reflected into the direction of observation by plural scattering[11] in the target must also be suppressed. If, for example, one makes a measurement at 120°, one does not find there only electrons that have been once scattered through 120°. One also finds electrons that have been scattered once through the angle α and then through $120° - \alpha$; or electrons that have reached the resulting angle 120° by more than two consecutive processes. The probability of such plural processes increases as the number of atoms in the target increases. This means that there are limits for the density of the Hg target and the thickness of the gold foil in

---

[11] Scattering in which more than one deflection takes place while not enough are involved to give the characteristic Gaussian distribution of multiple scattering.

the analyzer. For this reason, the density of the Hg beam in the experiment described had to correspond to a pressure which was considerably less than 10^{-3} Torr [3.19]. Likewise, the gold foil had an area density of $\sim 200\,\mu\text{g/cm}^2$, i.e., a thickness of approximately 1000 Å (cf. Subsect. 3.6.3).

3.6.2 Triple Scattering Experiments

By measuring the cross section in single scattering experiments and the Sherman function in double scattering experiments, one still has not found all the quantities that are required for a complete description of electron scattering. In addition, one needs the quantities $T(\theta)$ and $U(\theta)$ which describe the change of the polarization vector in a scattering process.

To measure these quantities one needs three consecutive scattering processes: In the first scattering the unpolarized electron beam becomes polarized. The second scattering process causes the change of the polarization vector, which is the object of the investigation. To be able to analyze this change of the polarization vector a third scattering is required—an asymmetry measurement with a Mott detector.

The problems that arise in double scattering become naturally much more pronounced in triple scattering. This is why only one measurement of $T(\theta)$ and $U(\theta)$ has so far been made [3.20]. To formally describe the experiment we resolve the polarization $\boldsymbol{P}$ before the crucial second scattering process into three mutually perpendicular components which refer to the direction of the beam before this scattering ($\hat{\boldsymbol{k}}_1$ = unit vector in this direction, see Fig. 3.26):

$$\boldsymbol{P} = (\boldsymbol{P}\cdot\hat{\boldsymbol{n}})\hat{\boldsymbol{n}} + (\boldsymbol{P}\cdot\hat{\boldsymbol{k}}_1)\hat{\boldsymbol{k}}_1 + (\boldsymbol{P}\cdot[\hat{\boldsymbol{k}}_1 \times \hat{\boldsymbol{n}}])[\hat{\boldsymbol{k}}_1 \times \hat{\boldsymbol{n}}]. \tag{3.88}$$

Fig. 3.26. Second scattering process in a triple scattering experiment

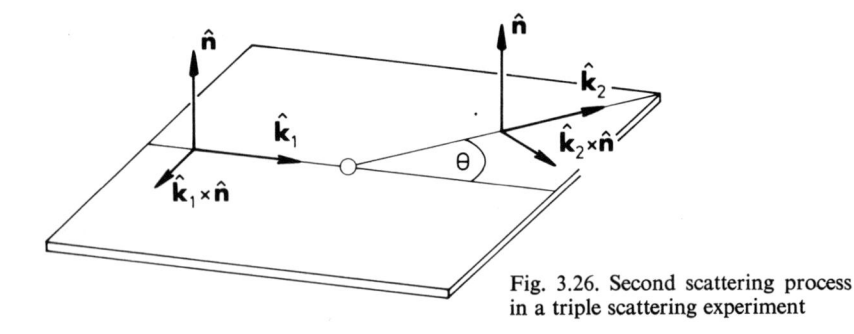

Similarly, the polarization after the second scattering process is resolved into three components which refer to the beam direction $\hat{\boldsymbol{k}}_2$ after

scattering. From (3.77) it then follows that (see Problem 3.7)

$$
\begin{aligned}
\boldsymbol{P}' = \frac{1}{1 + \boldsymbol{P} \cdot \hat{\boldsymbol{n}} S} \{ &\hat{\boldsymbol{n}}(\boldsymbol{P} \cdot \hat{\boldsymbol{n}} + S) + \hat{\boldsymbol{k}}_2(\boldsymbol{P} \cdot \hat{\boldsymbol{k}}_1 (T \cos \theta + U \sin \theta) \\
&+ \boldsymbol{P} \cdot [\hat{\boldsymbol{k}}_1 \times \hat{\boldsymbol{n}}](U \cos \theta - T \sin \theta)) \\
&+ [\hat{\boldsymbol{k}}_2 \times \hat{\boldsymbol{n}}](\boldsymbol{P} \cdot \hat{\boldsymbol{k}}_1 (T \sin \theta - U \cos \theta) \\
&+ \boldsymbol{P} \cdot [\hat{\boldsymbol{k}}_1 \times \hat{\boldsymbol{n}}](T \cos \theta + U \sin \theta)) \}.
\end{aligned} \qquad (3.89)
$$

Thus $\boldsymbol{P}'$ is resolved into components in the directions $\hat{\boldsymbol{n}}$, $\hat{\boldsymbol{k}}_2$ and $[\hat{\boldsymbol{k}}_2 \times \hat{\boldsymbol{n}}]$.

Since in this experiment the polarization $\boldsymbol{P}$ is produced by scattering, it cannot have a longitudinal component [see (3.73)], i.e., $\boldsymbol{P} \cdot \hat{\boldsymbol{k}}_1 = 0$. As the final polarization $\boldsymbol{P}'$ is measured with a Mott detector, its longitudinal component (direction $\hat{\boldsymbol{k}}_2$) cannot be detected. The quantity measured in the triple scattering experiment is thus

$$
\begin{aligned}
\boldsymbol{P}'_t = \frac{1}{1 + \boldsymbol{P} \cdot \hat{\boldsymbol{n}} S} \{ &\hat{\boldsymbol{n}}(\boldsymbol{P} \cdot \hat{\boldsymbol{n}} + S) \\
&+ [\hat{\boldsymbol{k}}_2 \times \hat{\boldsymbol{n}}](\boldsymbol{P} \cdot [\hat{\boldsymbol{k}}_1 \times \hat{\boldsymbol{n}}]) (T \cos \theta + U \sin \theta) \}.
\end{aligned} \qquad (3.90)
$$

Figure 3.27 is a schematic diagram of the apparatus used. The electrons were accelerated to 261 keV and focused by means of magnetic quadrupole lenses onto a gold foil in the first scattering chamber. The foil could withstand about 50 μA without deterioration. By sending the electrons that were scattered at 105° through an aperture, a transversely polarized beam with 30% polarization and a current of $1.5 \cdot 10^{-9}$ A was obtained.

If one wants to measure T and U, then according to (3.90) it is important to have a component of the polarization $\boldsymbol{P}$ in the direction $[\hat{\boldsymbol{k}}_1 \times \hat{\boldsymbol{n}}]$ which lies in the plane of the second scattering process. It would thus be expedient to choose the second scattering plane perpendicular to the first since $\boldsymbol{P}$ is normal to the first scattering plane. In practice it is simpler to also choose the second scattering plane as horizontal and to turn the polarization vector through a magnetic field into this plane (see Subsect. 3.6.3 on spin transformers). The magnetic lenses used for this simultaneously focus the beam onto the second scattering foil F. The polarization component $\boldsymbol{P}'_t$ arising after the scattering on this gold foil is analyzed at different scattering angles (in Fig. 3.27, the angle is 101°) by measuring the intensity asymmetry in the detectors I and II. Detectors III and IV are for correction of instrumental asymmetries (see next section).

Since the azimuthal direction ϕ of $\boldsymbol{P}'_t$ is not known, the Mott analyzer must be rotated about the final scattering direction to find the maximum

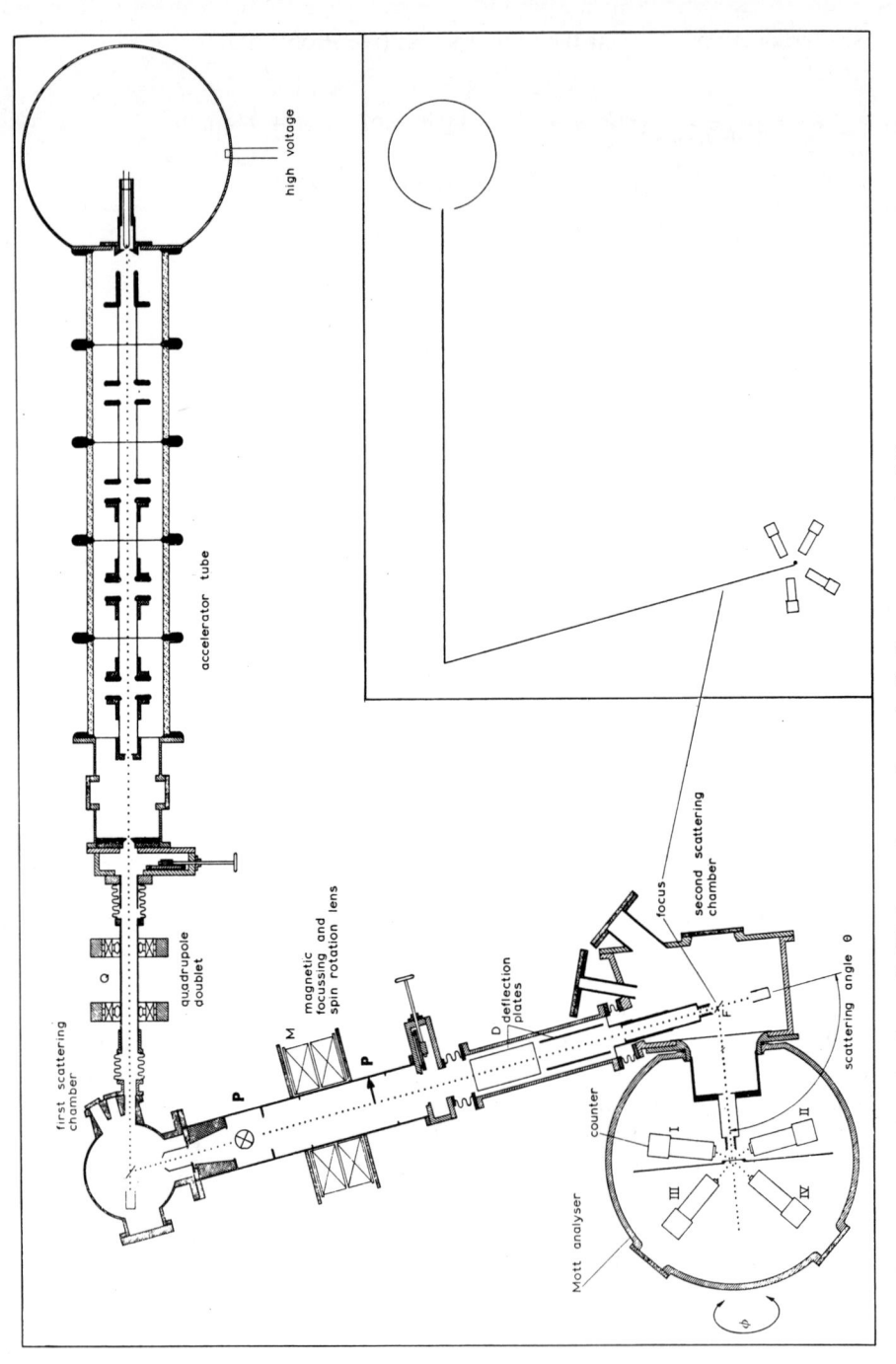

Fig. 3.27. Triple scattering experiment [3.20]. Insert shows schematically the measurement of *P*

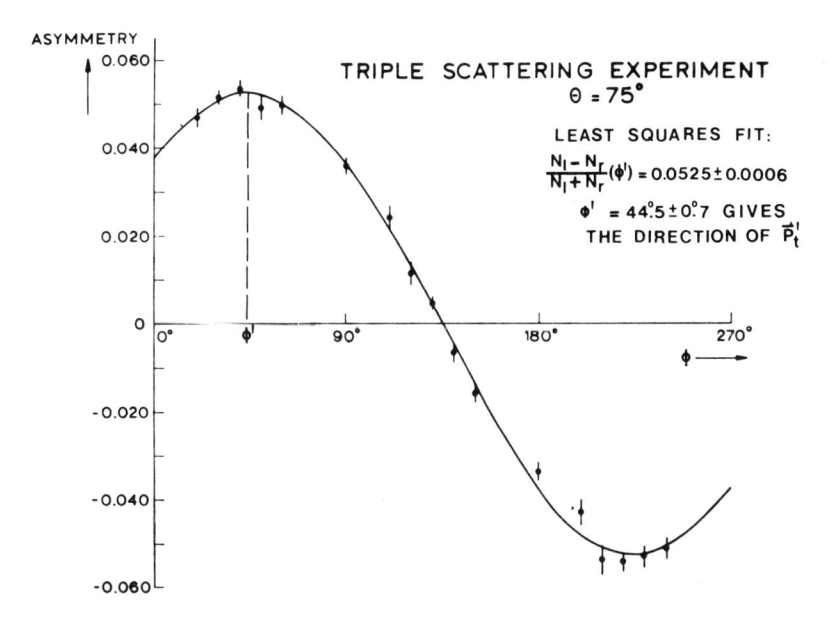

Fig. 3.28. Change of the intensity asymmetry with the rotation of the Mott chamber

left-right asymmetry. Figure 3.28 shows the cosine-shaped asymmetry curve for $\theta = 75°$ obtained in this way [cf. (3.70)].

When the Mott detector is in the position shown in Fig. 3.27 ($\phi = 0°$) the measurement yields only the component of $\boldsymbol{P}'_t$ parallel to $\hat{\boldsymbol{n}}$ since the Mott detector analyses only polarization components which lie perpendicular to the scattering plane it defines. Equation (3.90) shows that S can be determined from this component if the polarization vector $\boldsymbol{P}$ before the second scattering is known. $\boldsymbol{P}$ is measured by substituting the Mott detector for the second scattering foil, as shown schematically in the insert of Fig. 3.27.

For $\phi = 90°$ (position of Mott detector turned through $90°$ compared to the position shown in Fig. 3.27) the measurement yields the component of $\boldsymbol{P}'_t$ parallel to $[\hat{\boldsymbol{k}}_2 \times \hat{\boldsymbol{n}}]$; from this, according to (3.90), one obtains the quantity $T \cos \theta + U \sin \theta$.

With the form of the experiment chosen here, the term $T \sin \theta - U \cos \theta$ in (3.89) could not be measured, as it is no longer included in the measured quantity $\boldsymbol{P}'_t$ given by (3.90). Therefore the relation $S^2 + T^2 + U^2 = 1$ was used as a second independent equation for determining T and U. The ambiguity of the results due to this quadratic expression can be removed by referring to the theory. One has only to assume that the theory produces the correct sign of the quantity T. This is very likely because the theoretical prediction that T is seldom negative (see Fig. 3.23) corresponds to what one expects for physical reasons: $T < 0$ means a reversal of P_p, which rarely occurs since spin-orbit interaction is a weak force. It is therefore very unlikely that the signs given by the theory are wrong, since this would mean that T would be mostly negative and seldom positive.

The measurements were made at the angles $\theta = 45°$, $75°$, and $101°$. The final results of this difficult experiment naturally have rather large errors, but all the measured values lie near the calculated data of HOLZWARTH and MEISTER [3.16].

Problem 3.7: Resolve the polarization vector P' after the second scattering process into the components in the directions $\hat{n}$, $\hat{k}_2$, and $[\hat{k}_2 \times \hat{n}]$.

Solution: From (3.77) the polarization after scattering is
$$P' = \frac{[P \cdot \hat{n} + S(\theta)]\hat{n} + T(\theta)P_p + U(\theta)[\hat{n} \times P_p]}{1 + P \cdot \hat{n} S(\theta)}.$$

Thus the last two terms of the numerator of this expression must be transformed Using (3.88) one obtains for these terms
$$T\{(P \cdot \hat{k}_1)\hat{k}_1 + (P \cdot [\hat{k}_1 \times \hat{n}])[\hat{k}_1 \times \hat{n}]\}$$
$$+ U\{[\hat{n} \times \hat{k}_1(P \cdot \hat{k}_1)] + [\hat{n} \times [\hat{k}_1 \times \hat{n}](P \cdot [\hat{k}_1 \times \hat{n}])]\}.$$

As
$$\hat{k}_1 = (\hat{k}_1 \cdot \hat{k}_2)\hat{k}_2 + (\hat{k}_1 \cdot [\hat{k}_2 \times \hat{n}])[\hat{k}_2 \times \hat{n}] = \hat{k}_2 \cos\theta + [\hat{k}_2 \times \hat{n}] \sin\theta$$

one has
$$T(P \cdot \hat{k}_1)\{\hat{k}_2 \cos\theta + [\hat{k}_2 \times \hat{n}] \sin\theta\}$$
$$+ T(P \cdot [\hat{k}_1 \times \hat{n}])[\{\hat{k}_2 \cos\theta + [\hat{k}_2 \times \hat{n}] \sin\theta\} \times \hat{n}]$$
$$+ U(P \cdot \hat{k}_1)[\hat{n} \times \{\hat{k}_2 \cos\theta + [\hat{k}_2 \times \hat{n}] \sin\theta\}]$$
$$+ U(P \cdot [\hat{k}_1 \times \hat{n}])\{\hat{k}_2 \cos\theta + [\hat{k}_2 \times \hat{n}] \sin\theta\},$$

where $\hat{n} \times [\hat{k}_1 \times \hat{n}] = \hat{k}_1$ was used. Rearranging, substituting into the equation for P', and using $\hat{n} \times [\hat{k}_2 \times \hat{n}] = \hat{k}_2$ yields
$$P' = \frac{1}{1 + P \cdot \hat{n} S} \{\hat{n}(P \cdot \hat{n} + S) + \hat{k}_2(P \cdot \hat{k}_1(T \cos\theta + U \sin\theta)$$
$$+ P \cdot [\hat{k}_1 \times \hat{n}](U \cos\theta - T \sin\theta))$$
$$+ [\hat{k}_2 \times \hat{n}](P \cdot \hat{k}_1(T \sin\theta - U \cos\theta)$$
$$+ P \cdot [\hat{k}_1 \times \hat{n}](T \cos\theta + U \sin\theta))\}.$$

3.6.3 Experimental Equipment: Mott Detectors and Polarization Transformers

The principle of the polarization measurement with a Mott detector (determination of the left-right asymmetry) is very simple. In practice, however, there are many problems to overcome if the results are to be reliable.

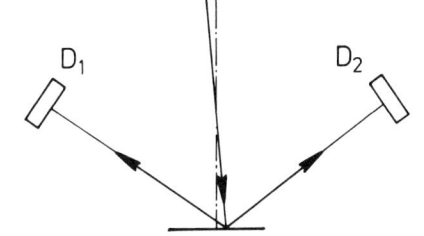

Fig. 3.29. Example of instrumental scattering asymmetry

The experimental setup usually possesses a purely instrumental asymmetry which can be due to different efficiencies of the detectors D_1 and D_2, not strictly axial alignment of the primary beam (see Fig. 3.29), or inhomogeneity of the target. This instrumental asymmetry must be determined and the measurements correspondingly corrected.

There are several methods for determining this asymmetry. One can, for example, reverse the polarization direction of the beam being investigated. In Fig. 3.25 this means, for example, rotating the electron gun to a position at the same angle θ symmetrical to its original position. This leads to reversal of the vector $\hat{n} = k_1 \times k_2/|k_1 \times k_2|$ normal to the scattering plane and thus of the polarization $P = S(\theta)\hat{n}$. Under ideal conditions the measured asymmetry should then also be reversed. The deviation from the "ideal" reversal of the asymmetry yields the extent of the instrumental asymmetry. One can also make a measurement with an unpolarized electron beam; then any asymmetry observed would be purely instrumental. In critical cases, it is useful if two more detectors, apart from D_1 and D_2, are set up symmetrically at small angles (see Fig. 3.27). Since at small angles the Sherman function always has a very small value near 0, no measurable asymmetry should occur here. In this way one can, for example, check the asymmetry due to non-axial beam alignment or due to unwanted changes in the beam that might arise from the polarization reversal just mentioned.

Since according to (3.70) or Fig. 3.1 the degree of polarization follows from the relation $(N_1 - N_r)/(N_1 + N_r) = PS(\theta)$, an exact knowledge of S is very important. One cannot simply use the theoretical value which was calculated for single scattering by one atom. Every real target contains so many atoms that plural or multiple scattering processes also occur; this becomes more likely, the thicker the target foil is (see end of Subsect. 3.6.1). Electrons which arrive at the counters after several consecutive scattering processes usually reduce the intensity asymmetry.

Consequently, instead of using the ideal Sherman function one must use an effective Sherman function which depends on the target thickness.

It also depends on other conditions of the experiment, for example, on the range of the scattering angle θ which is recorded by the electron detectors. Since there are rather large uncertainties in the theoretical treatment of plural scattering, it is advisable to experimentally ascertain the effective Sherman function for the Mott detector chosen. This can be done by calibrating the Mott detector with an electron beam of known polarization. It can also be done with a polarized beam of unknown polarization by measuring the increase of the scattering asymmetry with decreasing foil thickness d. If the amount of plural scattering is very small with the thinnest of the foils used, one can extrapolate from such a measured curve to the asymmetry for the foil thickness $d = 0$. This can be related to the Sherman function for $d = 0$ which is well established theoretically as well as experimentally for a suitable choice of scattering angle, electron energy, and target material (e.g., $\theta = 120°$, $E = 120$ keV, $Z = 79$). The measured curve then yields the effective Sherman function also for the other foils used in the series of measurements.

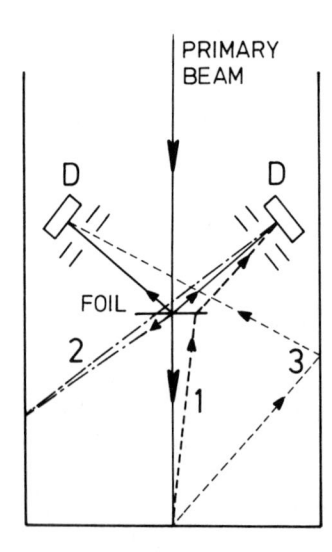

Fig. 3.30. Distortion of the measurement by backscattered electrons

Spurious electrons can arrive at the counters not only due to plural scattering in the foil but also due to reflection at the walls of the scattering chamber. Figure 3.30 shows a few typical cases:

1 Unscattered electrons from the incident beam are reflected on the chamber wall and after hitting the foil are scattered into a detector.

2 Scattered electrons hit the wall and are reflected into the direction of a detector.

3. Electrons from the incident beam reach a detector after double reflection at the wall.

In order to suppress these background electrons several of the following preventative measures are usually required: coating the inside of the chamber with a material having a small backscattering coefficient (small atomic number, e.g., carbon), suitable arrangement of diaphragms which capture the backscattered electrons, and good energy resolution of the counters since the electrons lose part of their energy due to the reflection.

To obtain significant left-right asymmetries, large values of the effective Sherman function S_{eff} are desirable. We have seen, however, in Subsection 3.4.2 that at angles where the Sherman function is large, the cross sections, and thus the scattering intensities, are small. Therefore one must find a compromise between high scattering intensity I and high asymmetry. It can easily be seen (see Problem 3.8) that one should choose $S_{eff}^2 I$ to be as large as possible, in order to make the statistical error as small as possible. As I also depends on the incident intensity I_0, it is reasonable to use the quantity $S_{eff}^2 I/I_0$ as a figure of merit when comparing different Mott detectors.

Since a detailed analysis shows that $S_{eff}^2 I$ increases with increasing foil thickness, one could come to the conclusion that it might be best to use a really thick scattering foil for the asymmetry measurement. However, one must not pay attention to the statistical error only. If the scattering foil is too thick, the effective value of the Sherman function and thus the left-right asymmetry will be very small. Then the systematic errors predominate —for example, those due to the instrumental asymmetry—and give rise to large error limits of the polarization measurement. An example of a favorable choice of foil thickness is the following: For electron energies of 120 keV and scattering angles of about 120°, gold foils with area densities between 0.1 and 0.5 mg/cm^2 (i.e., thicknesses of approximately 500–2500 Å) are suitable.

Even if one chooses a foil thickness in the upper range of the example just given, the values of I/I_0 still do not exceed a few times 10^{-3}. This means that from a thousand polarized electrons, only one can be detected. In comparison with this an analyzer for polarized light functions practically without loss. This is one of the reasons why experiments with polarized electrons are so much more difficult. For a Mott detector it is hardly possible to obtain an efficiency $S_{eff}^2 I/I_0$ which is much better than $5 \cdot 10^{-5}$ [3.21]. The corresponding value for an analyzer of polarized light, which, according to its position, almost totally absorbs or transmits polarized light is near 1. A Mott detector of this efficiency would have to scatter all the electrons of a totally polarized incident beam off to one side (see Fig. 3.31)!

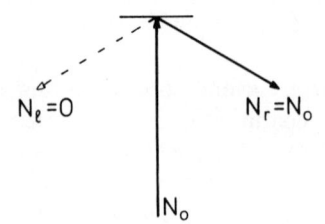

80   3. Polarization Effects in Electron Scattering from Unpolarized Targets

Fig. 3.31. Ideal Mott detector

Mott detectors can also of course be operated at low energies, for instance in the range 100–1000 eV. Then scattering foils can no longer be used because due to the much larger scattering cross sections at these energies there would be too much multiple scattering. One therefore uses Hg-vapor beams of moderate density [3.22]. The advantage of this method is that, when experimenting with slow polarized electrons, it is not necessary to accelerate them afterwards to higher energies in order to measure their polarization. The disadvantage is that the efficiency is 100 to 1000 times smaller than with the Mott detector which uses higher energies and gold foils.

It is true that the Mott detector responds only to transverse polarization components. All the same, longitudinal polarization components can be detected if one transforms them into the transverse direction before the electrons are scattered in the Mott detector. This can, for example, be done with an electrostatic field that rotates the velocity vectors of the electrons through 90° but does not affect their magnetic moments (see Fig. 3.32).

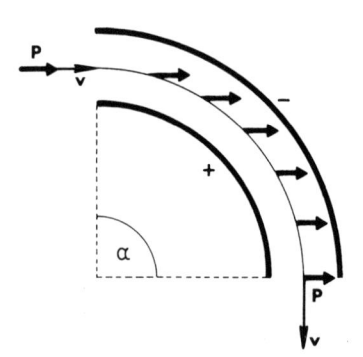

Fig. 3.32. Motion of a polarized electron beam in an electrostatic field

It is, however, only in the non-relativistic approximation that the electrostatic field does not affect the magnetic moments. To illustrate this we recall the picture used when discussing spin-orbit coupling: An electron experiences in its rest frame a magnetic field equal to $\boldsymbol{E} \times \boldsymbol{v}/c$. In the example of Fig. 3.32 this field is perpendicular to the plane of the diagram. Hence the spins precess in this field so that the polarization becomes slightly rotated when passing through the electric field. The exact calculation [3.23, 24] shows that the sector angle α of the electric field in Fig. 3.32 must be $\gamma\pi/2$ if the emerging polarization is to be transverse ($\gamma = 1/\sqrt{1 - (v/c)^2}$).

Needless to say, magnetic fields also can be used for spin rotation. This has been done, for example, in the triple scattering experiment in Subsection 3.6.2. Since a rigorous relativistic calculation [3.23, 24] of the spin rotation in arbitrary fields is complicated, we only discuss the most important special cases (a survey was given by FARAGO [3.25]). If the electrons move along the direction of the magnetic field and the polarization is perpendicular to it (see Fig. 3.33a), their direction of motion remains unchanged and $\boldsymbol{P}$ precesses about the field direction with a frequency which in the first approximation is $\omega = eB/m\gamma c$. The exact expression for the precession frequency for this particular case is

$$\omega = \frac{g}{2} \cdot \frac{eB}{m\gamma c}, \tag{3.91}$$

which is identical to the previous expression if $g = 2$. A more exact value of g is, however, $g = 2(1 + a)$ where $a = 1.16 \cdot 10^{-3}$ (see Sect. 7.2).

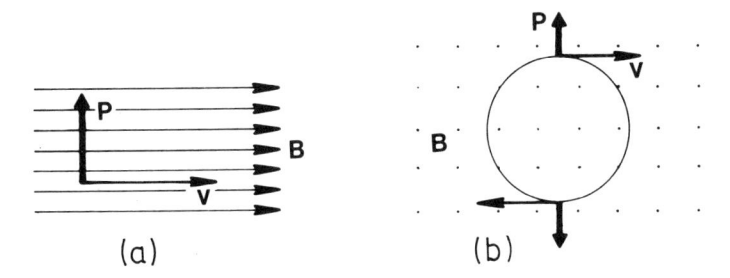

Fig. 3.33a and b. Motion of a transversely polarized electron beam (a) parallel and (b) perpendicular to a homogeneous magnetic field

We now consider the case of Fig. 3.33b where $\boldsymbol{B}$, $\boldsymbol{v}$ and $\boldsymbol{P}$ are mutually perpendicular to each other. Then the electrons move in a circular orbit with the cyclotron frequency

$$\omega_c = \frac{eB}{m\gamma c}. \tag{3.92}$$

If the g-factor were exactly 2, then precession frequency and cyclotron frequency would be equal so that the angle between $\mathbf{P}$ and $\mathbf{v}$ would remain unchanged. Since g is slightly larger than 2, the magnetic moment of the electrons is somewhat larger than the Bohr magneton so that $\mathbf{P}$ precesses slightly faster. The precession frequency differs by $a \cdot eB/mc$ from the value for $g = 2$

$$\omega_p = \frac{eB}{m\gamma c} + a \cdot \frac{eB}{mc} = \frac{eB}{mc}\left(\frac{1}{\gamma} + a\right). \tag{3.93}$$

We note only briefly that the difference from the frequency in (3.91) for motion along the magnetic field is due to the fact that with a circular motion, a Lorentz transformation does not lead directly to the rest frame but to a precessing frame (Thomas precession).

Due to the gradual advance of the polarization caused by $g > 2$, there is a gradual change in the angle between $\mathbf{P}$ and $\mathbf{v}$, so that after nearly 10^3 revolutions of the electrons the polarization vector has made one extra revolution (when $\gamma \approx 1$). This means that after approximately 200 revolutions, the polarization has been transformed from longitudinal to transverse (or vice versa).

Transformation of longitudinal into transverse polarization, or vice versa, is also possible without affecting the orbit. This happens in the Wien filter shown in Fig. 3.34. In the crossed electro- and magnetostatic fields, the electrons experience no resultant force if $(d/dt)(mv) = eE - eBv/c = 0$, i.e., $v = Ec/B$. Their spins, however, precess through the angle $eLB^2/m\gamma^2c^2E$ (L = length of the Wien filter), as can immediately be seen for the non-relativistic limiting case.

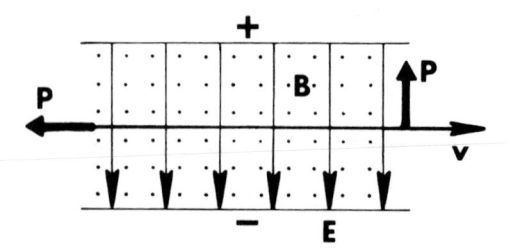

Fig. 3.34. Wien filter as a spin rotator

Problem 3.8: Calculate the statistical error of the polarization measurement with a Mott detector and establish the assertion that $S_{\text{eff}}^2 I$ should be as large as possible (if systematic errors are ignored).

Solution: Since $A = (N_1 - N_r)/(N_1 + N_r) = PS_{\text{eff}}$, the error of the polarization measurement for a given S_eff is

$$\Delta P = \frac{1}{S_\text{eff}} \cdot \Delta A.$$

From the law of the propagation of errors, the error ΔA of the measured asymmetry is expressed in terms of the errors of the individual measurements ΔN_l and ΔN_r by

$$\Delta A = \sqrt{\left(\frac{\partial A}{\partial N_l}\right)^2 (\Delta N_l)^2 + \left(\frac{\partial A}{\partial N_r}\right)^2 (\Delta N_r)^2}$$

$$= \sqrt{\left(\frac{2N_r}{(N_l + N_r)^2}\right)^2 N_l + \left(\frac{-2N_l}{(N_l + N_r)^2}\right)^2 N_r}$$

(for the errors ΔN_i of the individual measurements, the statistical errors $\sqrt{N_i}$ have been substituted). Setting $N_l + N_r = N$ one obtains

$$\Delta A = \sqrt{\frac{4N_r N_l}{N^3}}.$$

Since $1 - P^2 S_\text{eff}^2 = 4N_l N_r / N^2$, it follows that

$$\Delta A = \sqrt{\frac{1}{N}(1 - P^2 S_\text{eff}^2)}$$

and

$$\Delta P = \sqrt{\frac{1}{N}\left(\frac{1}{S_\text{eff}^2} - P^2\right)}.$$

With the Mott detectors used in practice, the effective Sherman functions are not very large so that $1/S_\text{eff}^2 > 10P^2$; thus

$$\Delta P = \sqrt{\frac{1}{NS_\text{eff}^2}}.$$

Since N, the number of particles observed, is (under otherwise identical conditions) proportional to the scattering intensity I, ΔP becomes smaller as $S_\text{eff}^2 I$ becomes larger.

3.7 Inelastic Scattering, Resonance Scattering, Electron-Molecule Scattering. Further Processes Used for Polarization Analysis

Spin polarization of the scattered electrons also occurs in inelastic scattering, resonance scattering, and electron-molecule scattering. To detect the polarization of electrons, one can—among other methods—also use the circular polarization of their bremsstrahlung or electron-electron scattering.

We have discussed in detail the important case of elastic scattering by the field of a single atom. The polarization effects arising in electron scattering are not, however, restricted to this case. We shall mention a few other examples.

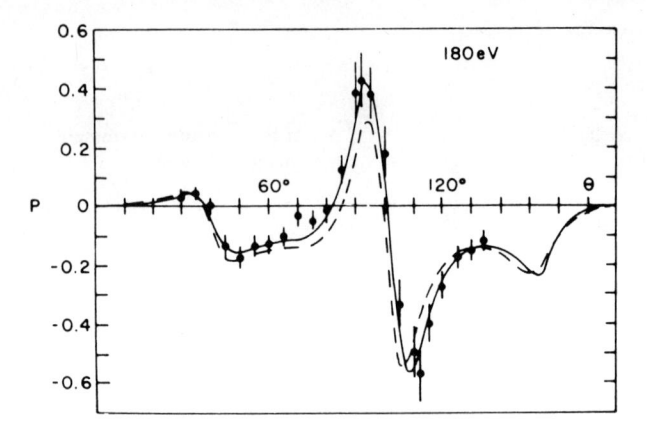

Fig. 3.35. Polarization of electrons scattered inelastically by Hg atoms (excitation of the 6^1P_1 state; energy loss 6.7 eV). Experimental [3.26] and theoretical [3.27] values

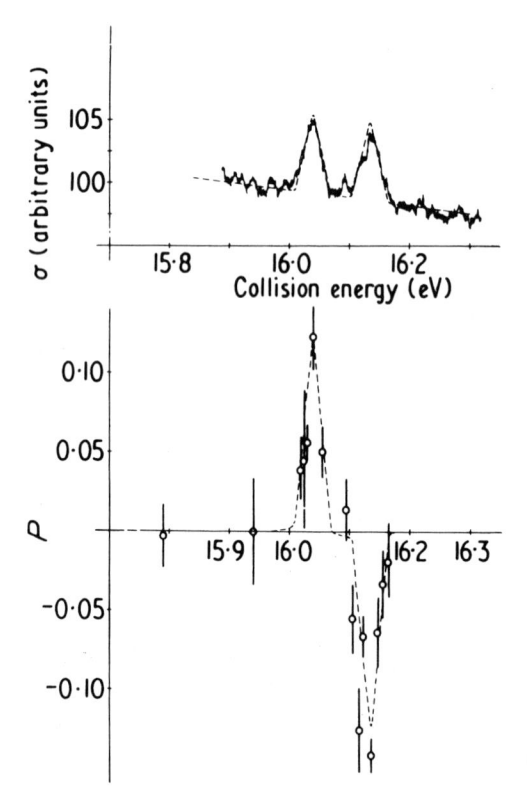

Fig. 3.36. Resonance scattering by neon, scattering angle $\theta = 90°$. Resonance structure of the differential cross section and corresponding polarization [3.29]

3.7 Inelastic, Resonance, Electron-Molecule Scattering   85

Spin polarization of the scattered electrons can also arise from inelastic scattering. In the past few years, experimental and theoretical investigations [3.26–28] have shown that electrons that excite a Hg atom and are thereby scattered through fairly large angles, are polarized. Figure 3.35 shows an example.

Polarization effects may also occur in resonance scattering of electrons by atoms where a compound state of a negative ion with a short lifetime is formed [3.29, 30]. Figure 3.36 shows that the resonances in the cross section can be correlated with the extrema of the polarization of the scattered electrons. Contrary to the Mott scattering discussed before, the polarization peaks are here connected with maxima of the cross sections.

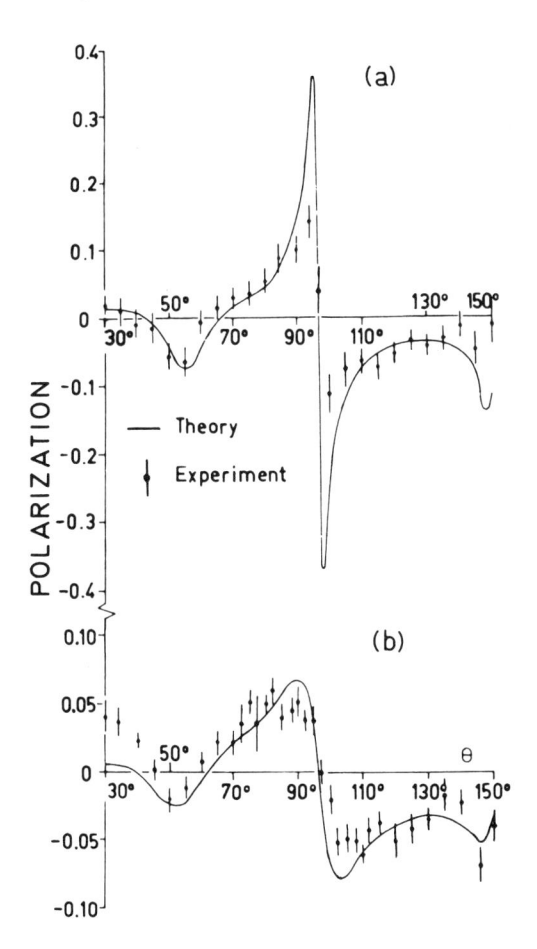

Fig. 3.37a and b. Experimental and calculated values of the polarization of 300-eV electrons after elastic scattering by (a) I_2 and (b) C_2H_5I molecules [3.31–33]

Since electrons can be polarized by scattering from atoms with fairly high atomic numbers, polarization effects are likewise to be expected in scattering by molecules that contain such atoms. These effects have in fact been observed (see Fig. 3.37). If, for example, a molecule contains light atoms as well as a heavy atom, the electrons scattered by the heavy atom are "marked" by their polarization; one can distinguish them from the unpolarized electrons which were scattered by the light atoms. From this, details about the scattering process and the molecular structure can be found [3.31–33] (see Sect. 7.1).

Although electron-atom scattering is certainly the most important method for measuring electron polarization, it is not the only one. For certain purposes, mainly of nuclear physics, there are other methods (see references given in [3.5]) from which we single out two that have been repeatedly used.

Longitudinally polarized electrons produce X-rays (bremsstrahlung) which have partial circular polarization. The polarization of the photons can be detected by Compton scattering or absorption in magnetized iron because the Compton cross section depends on the orientation of the photon spins with respect to the spin direction of the scattering electrons.

Another possibility for measuring electron polarization is electron-electron scattering (Møller scattering). Here, use is made of the fact that the cross section for the scattering of two electrons on each other depends on the spin direction of the electrons; the cross section is considerably smaller for parallel spins than for antiparallel spins. This difference can be visualized as follows: According to the Pauli principle, electrons with the same spin direction will on the average be further apart than those with opposing spin directions. They are thus less likely to be scattered by each other.[12] Accordingly, if one scatters polarized electrons by the electrons of a magnetized iron foil, whose spins are aligned first parallel and then antiparallel to the polarization direction of the incident electrons, one obtains different scattering intensities.

These remarks on electron-electron scattering lead us to the next chapter which is concerned with the influence of the spin of the target.

---

[12] "Exchange scattering", which is important here, will be treated in the next chapter for the more general case of atoms with bound electrons.

4. Exchange Processes in Electron-Atom Scattering

4.1 Polarization Effects in Elastic Exchange Scattering

If unpolarized electrons are elastically scattered by unpolarized atoms it is impossible to separate exchange scattering from direct scattering. This can, however, be done by using polarized electrons and/or atoms and observing the polarization of the scattered electrons or of the recoil atoms.

In the last chapter we saw that the spin-orbit interaction can give rise to a spin polarization in electron scattering. Another mechanism that can cause a polarization of the scattered electrons is the exchange interaction. Consider, for example, elastic scattering from a target of alkali atoms whose valence electrons all have the same spin direction. If exchange processes occur between the valence electrons and the free electrons, one obtains polarized electrons in the scattered beam.

In this section we deal with the elastic exchange scattering of electrons by atoms with one valence electron (hydrogen, alkalis). The polarization effects to be discussed are a consequence of the Pauli principle rather than of explicit spin-dependent forces between the colliding electrons: The dipole-dipole interaction between the incident and atomic electrons is small compared with the Coulomb interaction for all energies considered here; its influence on the scattering is completely masked by the influence of the Coulomb interaction. Furthermore we make the assumption that the spin-orbit interaction, which causes the polarization effects in Mott scattering, can be neglected. This is a good approximation for light atoms or small scattering angles (cf. the end of Sect. 4.2).

Let the scattering take place on an atom of arbitrary spin direction which we denote by A↑. For the scattering of electrons with spins parallel or antiparallel to the atomic spin the following possibilities are conceivable:

$$e\downarrow + A\uparrow \rightarrow e\downarrow + A\uparrow \qquad (4.1)$$

$$e\downarrow + A\uparrow \rightarrow e\uparrow + A\downarrow \qquad (4.2)$$

$$e\uparrow + A\uparrow \rightarrow e\uparrow + A\uparrow. \qquad (4.3)$$

4. Exchange Processes in Electron-Atom Scattering

In the first two processes the two colliding electrons can be distinguished: Since explicit spin-dependent forces were excluded, each electron retains its spin direction during the scattering process and is thus marked by this spin direction.

We recall from elementary scattering theory that the scattering amplitude can be expressed by

$$-\frac{m}{2\pi\hbar^2}\langle\psi_f|T|\psi_i\rangle, \tag{4.4}$$

where, in the first Born approximation, T is the scattering potential and ψ_i and ψ_f are the wave functions of the initial and the final state. In the scattering process discussed here, two electrons are involved: the incident electron and the valence electron. In order to describe these identical particles properly, one has to use antisymmetric wave functions in the above formula, thus taking the Pauli principle into account. We therefore write

$$\psi_i = \frac{1}{\sqrt{2}}[e^{i\mathbf{k}\mathbf{r}_1}\eta(1)u(\mathbf{r}_2)\chi(2) - e^{i\mathbf{k}\mathbf{r}_2}\eta(2)u(\mathbf{r}_1)\chi(1)] \tag{4.5a}$$

and

$$\psi_f = \frac{1}{\sqrt{2}}[e^{i\mathbf{k}'\mathbf{r}_1}\eta'(1)u'(\mathbf{r}_2)\chi'(2) - e^{i\mathbf{k}'\mathbf{r}_2}\eta'(2)u'(\mathbf{r}_1)\chi'(1)], \tag{4.5b}$$

where u and u' are the atomic wave functions in the initial and the final state; $\mathbf{k}$ and $\mathbf{k}'$ are the wave vectors of the incident and the scattered electrons; η, χ and η', χ' are the spin functions of the free and the bound electron in the initial and the final state, respectively.

With these wave functions we obtain the scattering amplitude

$$-\frac{m}{2\pi\hbar^2}\langle\psi_f|T|\psi_i\rangle = \frac{1}{2}\{f(\theta)[\langle\eta'(1)|\eta(1)\rangle\langle\chi'(2)|\chi(2)\rangle$$
$$+ \langle\eta'(2)|\eta(2)\rangle\langle\chi'(1)|\chi(1)\rangle]$$
$$- g(\theta)[\langle\eta'(1)|\chi(1)\rangle\langle\chi'(2)|\eta(2)\rangle$$
$$+ \langle\eta'(2)|\chi(2)\rangle\langle\chi'(1)|\eta(1)\rangle]\}, \tag{4.6}$$

where

$$f(\theta) = -\frac{m}{2\pi\hbar^2}\langle e^{i\mathbf{k}'\mathbf{r}_\lambda}u'(\mathbf{r}_\mu)|T|e^{i\mathbf{k}\mathbf{r}_\lambda}u(\mathbf{r}_\mu)\rangle \tag{4.7a}$$

$$g(\theta) = -\frac{m}{2\pi\hbar^2}\langle e^{i\mathbf{k}'\mathbf{r}_\mu}u'(\mathbf{r}_\lambda)|T|e^{i\mathbf{k}\mathbf{r}_\lambda}u(\mathbf{r}_\mu)\rangle \tag{4.7b}$$

4.1 Polarization Effects in Elastic Exchange Scattering

with $\lambda, \mu = 1, 2$ or $2, 1$. The scattering amplitude $g(\theta)$ obviously describes a process in which the incident electron is captured by the atom and the atomic electron is ejected. One therefore calls $g(\theta)$ the exchange amplitude, whereas $f(\theta)$ is called the direct amplitude. Needless to say, the exchange amplitude has nothing to do with the amplitude $g(\theta)$ which we introduced in the treatment of Mott scattering.

Let us describe the spin directions $\uparrow$ and $\downarrow$ in (4.1) to (4.3) by the respective spin functions $\alpha = \begin{pmatrix} 1 \\ 0 \end{pmatrix}$ and $\beta = \begin{pmatrix} 0 \\ 1 \end{pmatrix}$. Then we have for the process (4.1)

$$\eta = \beta, \quad \chi = \alpha, \quad \eta' = \beta, \quad \chi' = \alpha,$$

so that (4.6) yields the scattering amplitude

$$\tfrac{1}{2}\{f(\theta)[1 + 1] - g(\theta)\cdot 0\} = f(\theta).$$

In the process (4.2) we have $\eta = \beta, \chi = \alpha, \eta' = \alpha, \chi' = \beta$, resulting in the scattering amplitude $-g(\theta)$. In (4.3) the spin functions are $\eta = \chi = \eta' = \chi' = \alpha$, so that (4.6) yields the scattering amplitude $f - g$.

Since the cross sections are the squares of the scattering amplitudes we can summarize our results as follows:

Process	Cross section	
$e\downarrow + A\uparrow \to e\downarrow + A\uparrow$	$|f(\theta)|^2$	(4.8)
$e\downarrow + A\uparrow \to e\uparrow + A\downarrow$	$|g(\theta)|^2$	(4.9)
$e\uparrow + A\uparrow \to e\uparrow + A\uparrow$	$|f(\theta) - g(\theta)|^2$	(4.10)

These relations demonstrate the obvious physical meaning of the "direct" and the "exchange" cross section. The corresponding results for $A\downarrow$ are

$e\uparrow + A\downarrow \to e\uparrow + A\downarrow$	$|f(\theta)|^2$	(4.11)
$e\uparrow + A\downarrow \to e\downarrow + A\uparrow$	$|g(\theta)|^2$	(4.12)
$e\downarrow + A\downarrow \to e\downarrow + A\downarrow$	$|f(\theta) - g(\theta)|^2$	(4.13)

By scattering electrons and atoms with well-defined spin states on each other, and by analyzing the spin directions of the observed electrons and/or atoms, one could determine each of the cross sections listed above.

Needless to say, such experiments are very difficult. Usually unpolarized particles are used for scattering experiments and a spin analysis

4. Exchange Processes in Electron-Atom Scattering

after scattering is not made. In this case one measures the sum of the cross sections given above and the information on the individual contributions is lost. If, for example, one scatters an unpolarized electron beam by an A↑ target and observes all the scattered electrons independent of their spin direction, it follows from (4.8) to (4.10) that the cross section is

$$\frac{d\sigma}{d\Omega} \equiv \sigma(\theta) = \tfrac{1}{2}(|f(\theta)|^2 + |g(\theta)|^2) + \tfrac{1}{2}|f(\theta) - g(\theta)|^2$$
$$= \tfrac{1}{4}|f(\theta) + g(\theta)|^2 + \tfrac{3}{4}|f(\theta) - g(\theta)|^2. \tag{4.14}$$

The factor 1/2 is due to the fact that the cross sections given in (4.8) to (4.13) are valid for totally polarized beams; for an unpolarized primary beam, which may be considered to be made up of equal numbers of e↑ and e↓, the scattering intensities in the individual channels are only half as large as for a totally polarized beam. It is obvious that we would have obtained the same result if we had considered scattering of an unpolarized electron beam by an A↓ target (opposite spin direction). Thus the cross section (4.14) also describes the scattering of an unpolarized beam by an unpolarized target made up of equal numbers of A↑ and A↓.[1]

The validity of the last equality in (4.14) follows immediately from elementary rules for complex number calculation. The physical meaning of this form of the cross section can be seen as follows: When an unpolarized electron beam is scattered by an atom with one outer electron, the two colliding electrons may form either a triplet state $S = 1$ with the symmetric spin functions

$$\chi_S = \begin{cases} \alpha(1)\alpha(2) & \text{describing the substate } S_z = 1 \quad (4.15a) \\ \dfrac{1}{\sqrt{2}}[\alpha(1)\beta(2) + \beta(1)\alpha(2)] & \text{describing the substate } S_z = 0 \quad (4.15b) \\ \beta(1)\beta(2) & \text{describing the substate } S_z = -1 \quad (4.15c) \end{cases}$$

or a singlet state $S = 0$ with the antisymmetric spin function

$$\chi_A = \frac{1}{\sqrt{2}}[\alpha(1)\beta(2) - \beta(1)\alpha(2)]. \tag{4.16}$$

---

[1] For later purposes we state here that in this case neither the scattered electron beam nor the recoil atoms are, of course, polarized, since for every possible scattering process there exists an analogous process with opposite spins. Scattered electrons and atoms with each of the two spin directions occur with the same cross section $\sigma(\theta)/2$.

4.1 Polarization Effects in Elastic Exchange Scattering

The straightforward evaluation of the scattering amplitude (4.4) with the antisymmetric wave functions

$$\psi_i = \frac{1}{\sqrt{2}} [e^{i k r_1} u(r_2) \pm e^{i k r_2} u(r_1)] \chi_{A,S} \tag{4.17a}$$

$$\psi_f = \frac{1}{\sqrt{2}} [e^{i k' r_1} u'(r_2) \pm e^{i k' r_2} u'(r_1)] \chi_{A,S} \tag{4.17b}$$

yields $f - g$ for the three symmetric spin states and $f + g$ for the antisymmetric spin state. The differential cross section must be computed with the former term in three quarters of the collisions, and with the latter term in one quarter of the cases. We thus obtain the last expression of (4.14).

For the observation of the individual cross sections listed in (4.8) to (4.13) it is not necessary to use both polarized electrons and polarized atoms, as these equations may suggest.[2] It suffices to make a simpler experiment in which either the electrons or the atoms are initially unpolarized. If one, for example, scatters totally polarized electrons by unpolarized atoms then from (4.10) to (4.12) one has

Process		Cross section	
$e\uparrow + \begin{Bmatrix} A\uparrow \\ A\downarrow \end{Bmatrix} \rightarrow$	$\begin{cases} e\uparrow + A\uparrow \\ e\uparrow + A\downarrow \\ e\downarrow + A\uparrow \end{cases}$	$\frac{1}{2}\|f(\theta) - g(\theta)\|^2$	(4.18)
		$\frac{1}{2}\|f(\theta)\|^2$	(4.19)
		$\frac{1}{2}\|g(\theta)\|^2.$	(4.20)

Here, as in (4.14), it has been taken into account that for scattering by an unpolarized target (equal numbers of $A\uparrow$ and $A\downarrow$) the cross sections given in (4.10) to (4.12), which are valid for a totally polarized target, have to be multiplied by the factor $1/2$. As it is not possible to select scattered electrons of one spin direction for the analysis of these processes, one measures the polarization of the scattered electron beam

$$P'_e(\theta) = \frac{\sigma_e^\uparrow(\theta) - \sigma_e^\downarrow(\theta)}{\sigma_e^\uparrow(\theta) + \sigma_e^\downarrow(\theta)} = \frac{\frac{1}{2}|f(\theta) - g(\theta)|^2 + \frac{1}{2}|f(\theta)|^2 - \frac{1}{2}|g(\theta)|^2}{\frac{1}{2}|f(\theta) - g(\theta)|^2 + \frac{1}{2}|f(\theta)|^2 + \frac{1}{2}|g(\theta)|^2}. \tag{4.21}$$

Here $\sigma_e^\uparrow(\theta)$ and $\sigma_e^\downarrow(\theta)$ are the cross sections for the occurrence of $e\uparrow$ and $e\downarrow$, respectively, in the scattered beam. Their values have been taken from (4.18) to (4.20). According to (4.14), the denominator of (4.21) is none other than $\sigma(\theta)$, the cross section for the scattering of unpolarized electrons

---

[2] See, however, Problem 4.2.

by unpolarized atoms. If $\sigma(\theta)$ is known, the measurement of P'_e yields $\frac{1}{2}|f-g|^2 + \frac{1}{4}|f|^2 - \frac{1}{4}|g|^2$ and by subtraction of this quantity from $\sigma(\theta) = \frac{1}{2}|f-g|^2 + \frac{1}{4}|f|^2 + \frac{1}{4}|g|^2$ one obtains $|g|^2$.

If one observes in this experiment the recoil atoms $A\downarrow$, then according to (4.19) one obtains $|f|^2$. (Contrary to what we have discussed in Section 1.2 for the case of electrons, atoms of a certain spin direction can be selected, e.g., by a Stern-Gerlach type magnet; see also Sect. 4.2). With the help of three measurements (cross section for scattering of unpolarized particles, polarization of scattered electrons, fraction of the recoil atoms with certain spin direction) one thus can determine the quantities $|f|^2$, $|g|^2$, and $|f-g|^2$. Hence three of the four parameters of the complex quantities $f = |f|\exp(i\gamma_1)$ and $g = |g|\exp(i\gamma_2)$ can be determined: $|f|$ and $|g|$, and the difference of the phases of f and g from the relation $|f-g|^2 = |f|^2 + |g|^2 - 2|f||g|\cos(\gamma_1 - \gamma_2)$. One phase factor always remains arbitrary as has been explained in Subsection 3.3.3. (See also Problem 4.2).

In practice one cannot use totally polarized electron beams as we have assumed up to now. In principle this does not, however, change anything we have considered. A partially polarized electron beam with polarization P_e can be considered to be split up into two fractions in the ratio $P_e/(1-P_e)$ with total or zero polarizations, respectively (see Sect. 2.3). Accordingly, if we scatter a partially polarized electron beam by unpolarized atoms the cross section for occurrence of $e\uparrow$ in the scattered beam is

$$\sigma_e^\uparrow(\theta) = P_e[\tfrac{1}{2}|f(\theta) - g(\theta)|^2 + \tfrac{1}{2}|f(\theta)|^2] + (1-P_e)\frac{\sigma(\theta)}{2},$$

where $\sigma(\theta)$ is again the cross section for the scattering of unpolarized electrons by unpolarized atoms. Here use has been made of the fact that the unpolarized fraction of the incident electron beam produces equal numbers of $e\uparrow$ and $e\downarrow$ in the scattered beam (see footnote on p. 90). Correspondingly one has

$$\sigma_e^\downarrow(\theta) = P_e\tfrac{1}{2}|g(\theta)|^2 + (1-P_e)\frac{\sigma(\theta)}{2}.$$

Thus the polarization of the scattered electrons is

$$P'_e(\theta) = \frac{\sigma_e^\uparrow(\theta) - \sigma_e^\downarrow(\theta)}{\sigma_e^\uparrow(\theta) + \sigma_e^\downarrow(\theta)} = \frac{P_e}{\sigma(\theta)}[\sigma(\theta) - |g(\theta)|^2],$$

where use has been made of (4.14). One therefore obtains

$$|g(\theta)|^2 = \sigma(\theta)\left(1 - \frac{P'_e(\theta)}{P_e}\right). \tag{4.22}$$

4.1 Polarization Effects in Elastic Exchange Scattering

Hence, by a measurement of the electron polarization after the scattering, $|g(\theta)|$ can be determined if the electron polarization before the scattering and the cross section for the scattering of unpolarized electrons are known.

One can also observe the recoil atoms in this experiment. $A\uparrow$ emerge with the cross section

$$\sigma_A^\uparrow(\theta) = P_e[\tfrac{1}{2}|f(\theta) - g(\theta)|^2 + \tfrac{1}{2}|g(\theta)|^2] + (1 - P_e)\frac{\sigma(\theta)}{2}.$$

Similarly

$$\sigma_A^\downarrow(\theta) = P_e\tfrac{1}{2}|f(\theta)|^2 + (1 - P_e)\frac{\sigma(\theta)}{2}.$$

If we introduce the polarization of the atoms after the scattering

$$P_A'(\theta) = \frac{\sigma_A^\uparrow(\theta) - \sigma_A^\downarrow(\theta)}{\sigma_A^\uparrow(\theta) + \sigma_A^\downarrow(\theta)} = \frac{P_e}{\sigma(\theta)}[\sigma(\theta) - |f(\theta)|^2]$$

we obtain

$$|f(\theta)|^2 = \sigma(\theta)\left(1 - \frac{P_A'(\theta)}{P_e}\right), \tag{4.23}$$

which shows that a measurement of the polarization of the recoil atoms yields $|f(\theta)|$.

Thus we see that by scattering a partially polarized electron beam by unpolarized atoms all the individual cross sections listed in (4.8) to (4.10) can be obtained. Equivalent experiments can be made starting with unpolarized electrons and polarized atoms. In complete analogy to the above treatment, one then obtains

$$|f(\theta)|^2 = \sigma(\theta)\left(1 - \frac{P_e'(\theta)}{P_A}\right) \tag{4.24}$$

and

$$|g(\theta)|^2 = \sigma(\theta)\left(1 - \frac{P_A'(\theta)}{P_A}\right). \tag{4.25}$$

Hence one again obtains all the individual cross sections by measuring the electron polarization P_e' and the atomic polarization P_A' after scattering, if P_A, the polarization of the atoms before scattering, and $\sigma(\theta)$ are known.

We point out that the quantities $|f(\theta)|^2/\sigma(\theta)$ and $|g(\theta)|^2/\sigma(\theta)$ can actually be larger than 1. Example: if $g = f/2$ then from (4.14) one has $|f|^2/\sigma = 4/3$. This means that the polarization of the scattered particles in (4.22) to (4.25) can also be antiparallel to that of the particles which cause the polarization.

The emphasis of the present treatment was put on those cases in which either the primary beam or the target (but not both) is polarized, because such experiments are feasible today. Similar to the treatment of Mott scattering, the general case of scattering an electron beam of arbitrary direction of polarization on a polarized target can be easily obtained from the behavior of the two basic spin states, which was discussed at the beginning of this section. This is an application of the density-matrix formalism (see Problem 4.2) and shows that for a "perfect" scattering experiment, in which all observable quantities are measured, one must also investigate the rotation of the polarization component perpendicular to the target polarization.

In the next section a few quantitative results on elastic exchange scattering will be given. Additional discussions of the subjects of the present section can be found in [4.1–4].

Problem 4.1: In Subsections 3.3.2 and 3.4.5 it was proved that for Mott scattering one has $P = S$, where P is the polarizing power (for the scattering of an unpolarized beam) and S is the analyzing power (for the polarization measurement on a polarized beam). Are these two quantities also equal in exchange scattering? Compare the intensity asymmetries in exchange scattering and Mott scattering.

Solution: We first consider the case in which an unpolarized electron beam is polarized by scattering from totally polarized atoms A↑. From (4.9) and (4.10) one obtains scattered e↑ with the cross section $\frac{1}{2}|f-g|^2 + \frac{1}{2}|g|^2$, and from (4.8) scattered e↓ with cross section $\frac{1}{2}|f|^2$. Thus

$$P'_e = \frac{\frac{1}{2}|f-g|^2 + \frac{1}{2}|g|^2 - \frac{1}{2}|f|^2}{\frac{1}{2}|f-g|^2 + \frac{1}{2}|g|^2 + \frac{1}{2}|f|^2} = \frac{\sigma - |f|^2}{\sigma} = 1 - \frac{|f|^2}{\sigma}.$$

We now consider the scattering of a totally polarized electron beam by totally polarized atoms A↑. If the direction of polarization of the primary electrons is the same as that of the atoms, then according to (4.10) the scattering intensity is determined by the cross section $|f-g|^2$. According to (4.8) and (4.9), reversal of the polarization of the primary beam yields the cross section $|f|^2 + |g|^2$. The polarization analysis can be made by measuring the scattering intensity in a certain direction for the polarization directions ↑ and ↓ of the primary beam. The relative intensity difference gives us the analyzing power

$$S = \frac{|f-g|^2 - |f|^2 - |g|^2}{|f-g|^2 + |f|^2 + |g|^2} = 1 - \frac{|f|^2 + |g|^2}{\sigma},$$

which is different from the polarization calculated above.

The intensity asymmetries discussed here and those in Mott scattering are compared in Fig. 4.1. In exchange scattering of polarized electrons by polarized atoms there is, in the approximation made in this chapter, no left-right asymmetry since the forces do

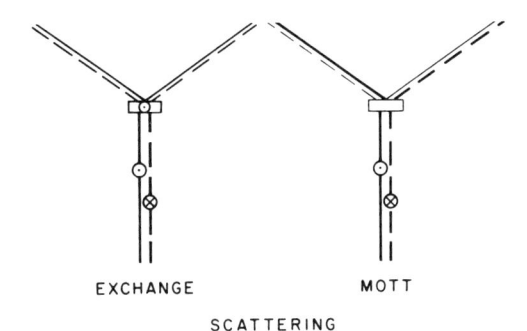

Fig. 4.1. Spin-up-down asymmetry and left-right asymmetry

not depend on whether an electron passes the atom on the right or on the left. However, the number of electrons scattered into a certain direction depends on their spin direction. So we do not have a "left-right" asymmetry, but we do have a "spin-up-down" asymmetry as illustrated in the left-hand part of Fig. 4.1. In Mott scattering there is a spin-up-down asymmetry as well as a left-right asymmetry.

4.2 Experiments on Polarization Effects in Elastic Exchange Scattering

Polarization experiments which have been carried out to measure the quantities $|f(\theta)|$ and $|g(\theta)|$ in elastic electron scattering are discussed. In these experiments one starts with polarized atoms and measures the polarization of either the recoil atoms or the scattered electrons.

In the last section it was seen that for atoms with one outer electron there are several experimental possibilities for determining the three cross sections that are essential for elastic scattering. The difficulties encountered with these experiments are of a diverse nature. For example, the scattering angles of the recoil atoms are only a few milliradians due to the large ratio of atomic masses to electron mass. On the other hand, polarized atoms can be produced or detected by means of suitable magnets without great loss of intensity, which is not true for electrons. In detecting the electron polarization (e.g., with a Mott detector) one suffers an intensity loss of several powers of ten; also the currents of polarized electrons which can be produced are several orders of magnitude less than the currents of unpolarized electrons (see Sect. 7.3).

The first experiments of this type used polarized atoms [4.5–8]. We shall first discuss an experiment [4.6] in which the change of the atomic spin by the scattering process is measured according to (4.25). In the second experiment [4.8] to be discussed the electron polarization P'_e after scattering is measured according to (4.24). This is already a double scattering experiment since one has the polarization analysis by scattering in addition to the scattering process which is to be studied. One even has a triple scattering experiment if, according to (4.22), one starts out with polarized electrons and analyzes the polarization P'_e of the scattered electrons: The first scattering yields polarized electrons with polarization P_e, the second scattering is the exchange process to be investigated, and the third analyzes P'_e.

Fig. 4.2. Schematic diagram of a scattering experiment with spin analysis of the recoil atoms [4.7]

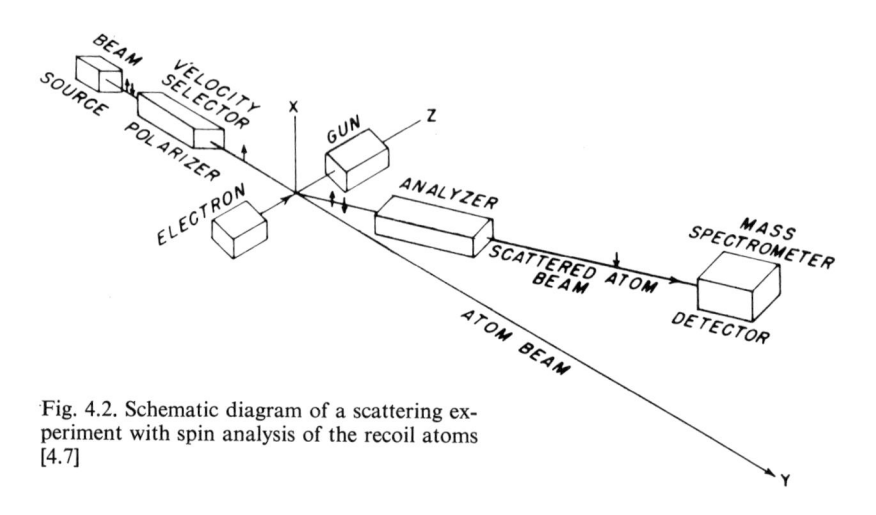

Figure 4.2 is a schematic diagram of the experiment in which a spin analysis of the recoil atoms was made. A beam of potassium atoms is sent through a Stern-Gerlach magnet which selects atoms of a certain spin direction and energy range. Slow electrons are fired across the polarized atom beam. Their energy is in the 1-eV region and is thus insufficient to excite the K atoms, so that only elastic collisions between the K atoms and the electrons occur. Both collision partners change their momentum direction during the scattering, the scattering angle of the heavy atoms being naturally much smaller ($\lesssim 1°$) than that of the electrons.

Due to the exchange interaction, some of the recoil atoms change their spin direction during the scattering process. They are selected by the spin analyzer if it is set so that only atoms with reversed spin direction are transmitted. This spin filter consists of a strong inhomogeneous magnetic field on which a strong inhomogeneous electric field is superimposed. The electric field is such that the force exerted on the induced electric dipole moments of the atoms has the same magnitude but the opposite direction as the magnetic force on the atoms with the desired spin direc-

4.2 Experiments on Polarization Effects in Elastic Exchange Scattering

tion. These atoms therefore pass through the analyzer without being deflected, while the atoms with the opposite spin direction are removed from the beam.

By rotating the analyzer-detector assembly about the scattering center, the angular distribution of the scattered atoms can be measured. From this, the angular distribution of the scattered electrons can be calculated using energy-momentum conservation.

If the spin filter is switched off, all atoms scattered within the angular range subtended by the detector are observed and not just those which have experienced a spin-flip. In this case one obtains the full differential cross section $\sigma(\theta)$. In the present experiment the ratio $|g(\theta)|^2/\sigma(\theta)$ was first measured by using the two analyzer settings mentioned. Then, in order to find $|g(\theta)|^2$, the quantity $Q = 2\pi\int\sigma(\theta)\sin\theta d\theta$ was determined from the measured angular distribution and fitted to the total cross section Q measured in a separate experiment. This yields the absolute values of $\sigma(\theta)$ and thus of $|g(\theta)|^2$.

Needless to say, there were deviations from the ideal conditions described here. For example, the imperfect transmission and polarization of analyzer and polarizer had to be determined by test measurements. The influence of the nuclear spins on the atomic polarization was avoided by using an additional strong magnetic field in the scattering region (see Sect. 5.1). The two spin systems were thereby decoupled so that the change of the atomic polarization may be assumed to be due solely to the scattering. The detection of the weak atomic-beam intensities was accomplished by using the lock-in technique: The electron beam was modulated with a certain frequency. This frequency was superimposed by the scattering process on the recoil atoms which reached the detector. The phase-sensitive narrow-band amplifier for the detector signal was locked to the oscillator controlling the electron beam modulation and therefore responded only to the modulation frequency. In this way all disturbing background frequencies were cut out.

The main difference between this measurement and the second experiment to be mentioned [4.8] is that instead of the spins of the recoil atoms, those of their collision partners, the scattered electrons, are analyzed. This yields $|f(\theta)|^2/\sigma(\theta)$ according to (4.24). The electrons scattered by the polarized potassium atoms pass through a filter lens which removes unwanted inelastically scattered electrons. They are then accelerated to 100 keV so that their polarization can be measured with a Mott detector, just as in the experiments described in Section 3.6.

In this experiment it was not possible to decouple electron and nuclear spins in the scattering region by using a strong magnetic field. This would have had too strong an effect on the path of the electrons whose scattering

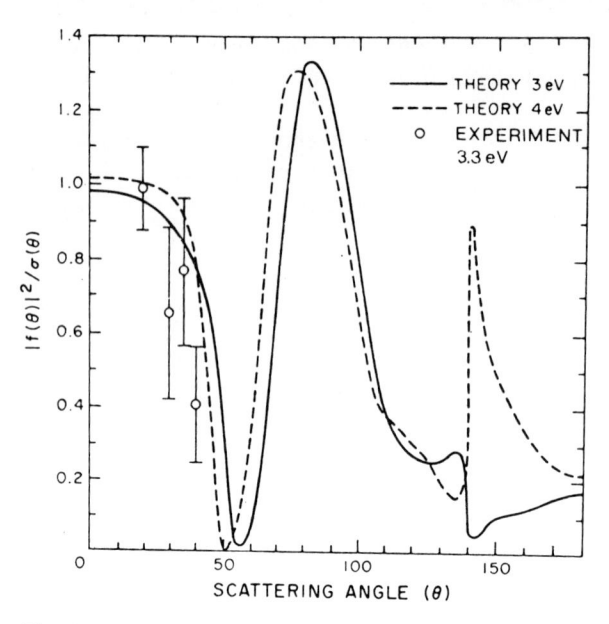

Fig. 4.3. Experimental [4.8] and theoretical [4.9] values of $|f(\theta)|^2/\sigma(\theta)$ for potassium

angle is one of the quantities to be measured. Accordingly, the polarization of the atoms is reduced by the hyperfine coupling to rather small values (<20%). In turn, the electron polarization is small, so that very low scattering asymmetries had to be measured. Furthermore, the experiment suffered from the lack of intensity, which is typical of many double scattering experiments.

These difficulties explain the rather large error bars on the results shown in Fig. 4.3. The measurements could be carried out only at relatively small scattering angles because only there were the cross sections large enough to produce usable intensities. Despite all the problems, the measurements show that calculations made by Russian authors [4.9] with the close-coupling approximation appear to be quite reliable. This is particularly brought out in the first experiment discussed where $|g(\theta)|^2$ was determined in a large angular range at several energies with an error of <30% (see Fig. 4.4).

The rather large cross sections $|g(\theta)|^2$ of more than 10^{-16} cm^2/sr suggest the use of exchange scattering of slow electrons for building a source of polarized electrons. We shall treat this and give further references in Section 7.3.

The evaluation of the above measurements has been made under the assumption that the role of spin-orbit coupling can be neglected. When one is able to increase the accuracy of the experiments, it will be necessary

4.2 Experiments on Polarization Effects in Elastic Exchange Scattering

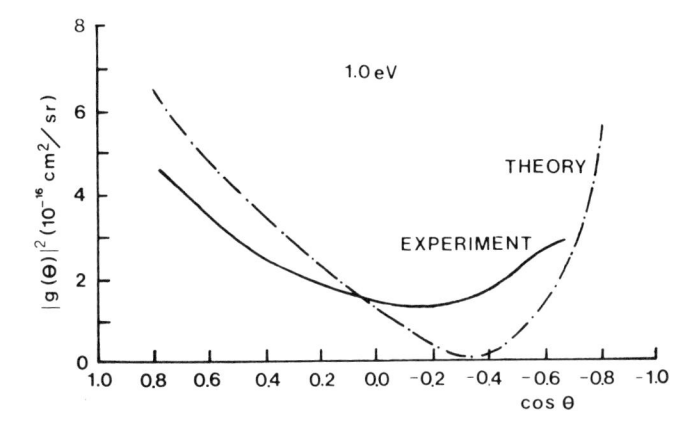

Fig. 4.4. Experimental [4.7] and theoretical [4.9] values of $|g|^2$ for potassium

to consider, quantitatively, how far the polarization effects can be influenced by spin-orbit interaction, particularly if heavier atoms and fairly large scattering angles are studied. The theoretical treatment of this problem has been initiated [4.10–12].

Problem 4.2: A totally polarized electron beam of arbitrary spin direction is scattered by an A↑ target. Calculate the polarization P'_e of the scattered electrons using the density-matrix formalism of Section 2.3.

Solution: When the z direction is defined by the spin direction of the atoms the scattering processes (4.1) to (4.3) can be described mathematically by the transformations

$$\begin{pmatrix} 0 \\ f \end{pmatrix} = \begin{pmatrix} 0 & 0 \\ 0 & f \end{pmatrix} \begin{pmatrix} 0 \\ 1 \end{pmatrix}, \quad \begin{pmatrix} g \\ 0 \end{pmatrix} = \begin{pmatrix} 0 & g \\ 0 & 0 \end{pmatrix} \begin{pmatrix} 0 \\ 1 \end{pmatrix}, \quad \begin{pmatrix} f-g \\ 0 \end{pmatrix} = \begin{pmatrix} f-g & 0 \\ 0 & 0 \end{pmatrix} \begin{pmatrix} 1 \\ 0 \end{pmatrix}.$$

If an electron beam of polarization $P = (P_x, P_y, P_z)$ is scattered by A↑ we therefore obtain for the density matrix of the scattered electrons

$$\rho' = \tfrac{1}{2} \begin{pmatrix} f-g & 0 \\ 0 & f \end{pmatrix} \begin{pmatrix} 1+P_z & P_x - iP_y \\ P_x + iP_z & 1-P_z \end{pmatrix} \begin{pmatrix} f^* - g^* & 0 \\ 0 & f^* \end{pmatrix}$$
$$+ \tfrac{1}{2} \begin{pmatrix} 0 & g \\ 0 & 0 \end{pmatrix} \begin{pmatrix} 1+P_z & P_x - iP_y \\ P_x + iP_y & 1-P_z \end{pmatrix} \begin{pmatrix} 0 & 0 \\ g^* & 0 \end{pmatrix}.$$

It has been taken into account here that due to process (4.2) which implies a transition to a different atomic spin state, a partially polarized electron beam (mixed state) is produced from a totally polarized beam of arbitrary spin direction; this process therefore yields an incoherent contribution. Straightforward evaluation yields

$$\rho' = \tfrac{1}{2} \begin{pmatrix} (1+P_z)|f-g|^2 + (1-P_z)|g|^2 & (P_x - iP_y)f^*(f-g) \\ (P_x + iP_y)f(f^* - g^*) & (1-P_z)|f|^2 \end{pmatrix}$$

and

$$\sigma(\theta) = \frac{\mathrm{tr}\,\rho'}{\mathrm{tr}\,\rho} = \tfrac{1}{2}[(1+P_z)|f-g|^2 + (1-P_z)(|f|^2 + |g|^2)], \quad (4.26)$$

$$P'_e = \frac{\text{tr}\, \rho'\sigma}{\text{tr}\, \rho'} = \frac{1}{2\sigma(\theta)} \{[(1 + P_z)|f - g|^2 + (1 - P_z)(|g|^2 - |f|^2)]\hat{e}_z$$
$$+ [2P_x|f|^2 - P_x(f^*g + fg^*) + iP_y(f^*g - fg^*)]\hat{e}_x$$
$$+ [2P_y|f|^2 + iP_x(fg^* - f^*g) - P_y(fg^* + f^*g)]\hat{e}_y\}. \tag{4.27}$$

This shows that the electron-polarization component perpendicular to the target polarization is rotated in the x-y plane. This rotation allows the quantities $fg^* + f^*g = 2\,\text{Re}\{fg^*\} = 2|f||g|\cos(\gamma_1 - \gamma_2)$ and $i(fg^* - f^*g) = -2\,\text{Im}\{fg^*\} = -2|f||g|\sin(\gamma_1 - \gamma_2)$ to be measured so that the phase difference $\gamma_1 - \gamma_2$ can be unambiguously determined. For a "perfect" scattering experiment in which all the observable quantities are measured an investigation of the rotation is necessary (as in Mott scattering, Subsect. 3.3.3); the measurements discussed in Section 4.1 are not quite sufficient, since knowledge of $\cos(\gamma_1 - \gamma_2)$ still leaves an ambiguity of the phase difference.

4.3 Polarization Effects in Inelastic Exchange Scattering

Excitation of atoms by polarized electrons is discussed. The cross sections for excitation of the various sublevels of one-electron atoms can be determined by analysis of the polarization of the inelastically scattered electrons in conjunction with measurement of the circular polarization of the emitted light. A triple scattering experiment on excitation of two-electron atoms by polarized electrons is discussed.

4.3.1 One-Electron Atoms

In elastic scattering, the atom remains in the ground state and at most changes its spin direction. In inelastic scattering, the atom is excited to other states which differ in their energies and their angular momenta. A separation of transitions into different angular momentum states cannot be achieved by energy analysis if the states are energetically degenerate; it can, however, be performed by the methods discussed below.

In order to describe the various inelastic transitions one needs more scattering amplitudes than in the elastic case, but there are also more observable quantities. A direct spin analysis of the excited atoms is not easy because their lifetimes are generally short. One can, however, obtain information on inelastic exchange scattering by analyzing the circular polarization of the light produced in the decay of the various excited states. Furthermore, the polarization of the atoms after their return to the ground state and, as in the elastic case, the polarization of the scattered electrons are important sources of information.

In line with our major topic, we shall mainly discuss the electron polarization and also deal with the polarization of the emitted light which arises after excitation by polarized electrons. It should, however, be mentioned that an experimental polarization analysis of the atoms that have returned to the ground state has also been carried out [4.13, 14]. A direct spin analysis of the excited atoms has been made for metastable hydrogen which has a sufficiently long lifetime [4.15].

In this subsection, we treat the simplest case of atoms with one outer electron and consider excitations from the ground state into the resonance states, that is, transitions $S \rightarrow P$. As in the case of elastic exchange scattering, we neglect the spin-orbit interaction of the unbound electrons that causes the polarization effects in Mott scattering.

For the P state there are, with respect to some reference direction, three possible orientations, specified by $m_l = 0, \pm 1$. Accordingly, three scattering amplitudes $f_0(\theta)$, $f_1(\theta)$, and $f_{-1}(\theta)$ are needed to describe the inelastic direct scattering leading to the excitation of one of these states. Correspondingly, three scattering amplitudes $g_0(\theta)$, $g_1(\theta)$, and $g_{-1}(\theta)$ are needed to describe the exchange scattering. From symmetry considerations, one has $|f_1|^2 = |f_{-1}|^2$ and $|g_1|^2 = |g_{-1}|^2$. If this were not the case, the two states $m_l = \pm 1$ would obtain different populations even when excited by ordinary collisions with unpolarized collision partners. The resulting angular momentum in the direction of quantization, together with the components in this direction of the momenta occurring in the scattering process, would define a screw-sense which would be reversed in the mirror image of the experiment. Thus—as explained in more detail in Subsection 3.4.4—parity conservation would be violated.

Applying to the inelastic case the results summarized in (4.8) to (4.13), we obtain

Process	Cross section	
$e\downarrow + A\uparrow \rightarrow e\downarrow + A(^2P)\uparrow$ $e\uparrow + A\downarrow \rightarrow e\uparrow + A(^2P)\downarrow$	$\begin{aligned} &\|f_0(\theta)\|^2 + \|f_1(\theta)\|^2 + \|f_{-1}(\theta)\|^2 \\ &= \|f_0(\theta)\|^2 + 2\|f_1(\theta)\|^2 \end{aligned}$	(4.28)
$e\downarrow + A\uparrow \rightarrow e\uparrow + A(^2P)\downarrow$ $e\uparrow + A\downarrow \rightarrow e\downarrow + A(^2P)\uparrow$	$\|g_0(\theta)\|^2 + 2\|g_1(\theta)\|^2$	(4.29)
$e\uparrow + A\uparrow \rightarrow e\uparrow + A(^2P)\uparrow$ $e\downarrow + A\downarrow \rightarrow e\downarrow + A(^2P)\downarrow$	$\|f_0(\theta) - g_0(\theta)\|^2 + 2\|f_1(\theta) - g_1(\theta)\|^2$,	(4.30)

where $A(^2P)$ represents an atom in the excited ^2P state and the tabulated cross section applies to each of the individual processes to its left. Analogous to the elastic case of (4.14), the differential cross section for the case where at least one of the colliding beams is unpolarized is thus

$$\sigma(^2P) = \tfrac{1}{2}|f_0|^2 + |f_1|^2 + \tfrac{1}{2}|g_0|^2 + |g_1|^2 + \tfrac{1}{2}|f_0 - g_0|^2 + |f_1 - g_1|^2. \quad (4.31)$$

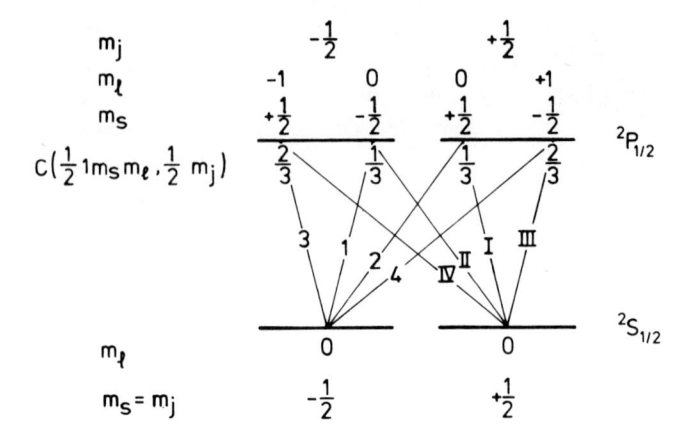

102   4. Exchange Processes in Electron-Atom Scattering

m_j $-\frac{1}{2}$ $+\frac{1}{2}$
m_ℓ -1 0 0 $+1$
m_s $+\frac{1}{2}$ $-\frac{1}{2}$ $+\frac{1}{2}$ $-\frac{1}{2}$ $^2P_{1/2}$
$C(\frac{1}{2} 1 m_s m_\ell, \frac{1}{2} m_j)$ $\sqrt{\frac{2}{3}}$ $\sqrt{\frac{1}{3}}$ $\sqrt{\frac{1}{3}}$ $\sqrt{\frac{2}{3}}$

                                                                                     $^2S_{1/2}$
m_ℓ 0 0
$m_s = m_j$ $-\frac{1}{2}$ $+\frac{1}{2}$

Transition		Cross section	Spin direction of scattered electron		Circular polarization of light emitted in transitions $\Delta m_j = +1$ or -1	
a)	b)		a)	b)	a)	b)
I	1	$\frac{1}{3}\|f_0 - g_0\|^2$	↑	↓	σ^+	σ^-
II	2	0				
III	3	0				
IV	4	$\frac{2}{3}\|f_1 - g_1\|^2$	↑	↓	σ^-	σ^+
c)	d)		c)	d)	c)	d)
I	1	$\frac{1}{3}\|f_0\|^3$	↓	↑	σ^+	σ^-
II	2	$\frac{1}{3}\|g_0\|^2$	↑	↓	σ^-	σ^+
III	3	$\frac{2}{3}\|g_1\|^2$	↑	↓	σ^+	σ^-
IV	4	$\frac{2}{3}\|f_1\|^2$	↓	↑	σ^-	σ^+

Fig. 4.5. Transitions to $^2P_{1/2}$ with totally polarized collision partners. a) to d) denote the following processes:

a) $e\uparrow + A\uparrow \to A(^2P_{1/2})$ b) $e\downarrow + A\downarrow \to A(^2P_{1/2})$
c) $e\downarrow + A\uparrow \to A(^2P_{1/2})$ d) $e\uparrow + A\downarrow \to A(^2P_{1/2})$

For later discussion of the light emission we have to take the fine structure of the P levels into account. The transitions to $^2P_{1/2}$ which occur with various initial polarizations are illustrated by Fig. 4.5. In characterizing the excited states by the quantum numbers m_s and m_l we tacitly assumed that the excitation time is short compared with the spin-orbit relaxation time so that the spin-orbit-coupled states need not be considered during excitation. This is justified for light alkali atoms, where the relaxation time is of order 10^{-12} s, whereas the excitation time may be considered to be of order 10^{-15} s, if the energy spread of the electron wave packet is not extraordinarily small. For all processes that occur after a time larger than $\sim 10^{-12}$ s, we have to take into account the fact that due to spin-orbit coupling m_s and m_l are not good quantum numbers; from the quantum numbers of the excited states in Fig. 4.5, only j and m_j represent constants of the motion.

Using the Clebsch-Gordan coefficients $C(slm_sm_l, jm_j)$ tabulated in textbooks on quantum mechanics, a state with certain values j, m_j can be written as a superposition of states[3] $|m_s, m_l\rangle$ which have fixed values of m_s and m_l, where $m_s + m_l = m_j$. For example, the wave function for the $m_j = \frac{1}{2}$ substate of $^2P_{1/2}$ is given by[4]

$$\sqrt{\tfrac{1}{3}}|\tfrac{1}{2}, 0\rangle - \sqrt{\tfrac{2}{3}}|-\tfrac{1}{2}, 1\rangle, \tag{4.32}$$

which is a superposition of the wave functions corresponding to $m_s = +\frac{1}{2}$, $m_l = 0$, and $m_s = -\frac{1}{2}$, $m_l = 1$. The Clebsch-Gordan coefficients $\sqrt{\tfrac{1}{3}}$ and $-\sqrt{\tfrac{2}{3}}$ are chosen so that (4.32) is an eigenfunction not only of j_z but also of j^2, since the total angular momentum also is a constant of the motion (see also Subsect. 5.2.1 and Problem 5.1).

We recall from elementary quantum mechanics that in an expansion $|\psi\rangle = \sum_n c_n|u_n\rangle$, where the $|u_n\rangle$ obey the equation $Q|u_n\rangle = q_n|u_n\rangle$, $|c_n|^2$ represents the probability of finding the eigenvalue q_n when a measurement of the observable Q is made on a system described by ψ. Applied to our specific case, this means that the square of the Clebsch-Gordan coefficient with which a state $|m_s, m_l\rangle$ is multiplied gives the probability of finding the atom in this state. In other words, the squares of these coefficients describe the contributions of the individual uncoupled states to the coupled state. Hence, the cross section for a transition into the coupled state $|j, m_j\rangle$ is found by adding the cross sections for reaching the uncoupled states from which $|j, m_j\rangle$ is linearly combined, after multiplying with the squares of the corresponding Clebsch-Gordan coefficients. These squares are given for the individual states in the diagram of Fig. 4.5 and as numerical factors of the cross sections in the figure caption. The last column concerns the polarization of emitted radiation, as will be explained later.

The transitions into the $^2P_{3/2}$ levels are illustrated by Fig. 4.6 in the same way.

We now consider the experimental possibilities for determining the quantities introduced here and choose as an example the excitation of unpolarized atoms by polarized electrons e↑ with polarization P_e. Let us assume that the quantity to be measured is the polarization P'_e of the scattered electrons after the excitation process. Then we do not have to

---

[3] Since we consider states of fixed s and l here, we dispense with these quantum numbers in the state vectors.
[4] As to the sign of the Clebsch-Gordan coefficients, cf. CONDON-SHORTLEY [4.16, p. 123].

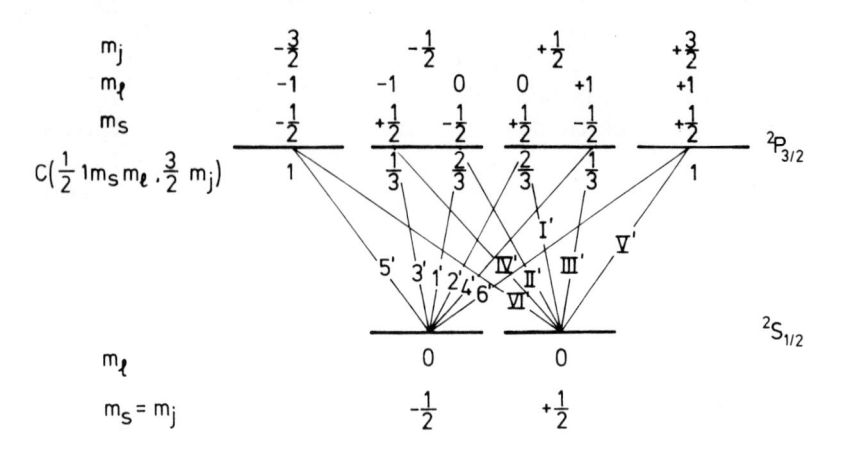

4. Exchange Processes in Electron-Atom Scattering

	$-\frac{3}{2}$	$-\frac{1}{2}$		$+\frac{1}{2}$		$+\frac{3}{2}$	
m_j							
m_ℓ	-1	-1	0	0	$+1$	$+1$	
m_s	$-\frac{1}{2}$	$+\frac{1}{2}$	$-\frac{1}{2}$	$+\frac{1}{2}$	$-\frac{1}{2}$	$+\frac{1}{2}$	${}^2P_{3/2}$
$C(\frac{1}{2} 1 m_s m_\ell, \frac{3}{2} m_j)$	1	$\frac{1}{\sqrt{3}}$	$\sqrt{\frac{2}{3}}$	$\sqrt{\frac{2}{3}}$	$\frac{1}{\sqrt{3}}$	1	

${}^2S_{1/2}$

m_ℓ	0	0
$m_s = m_j$	$-\frac{1}{2}$	$+\frac{1}{2}$

Transition		Cross section	Spin direction of scattered electron		Circular polarization of light emitted in transitions $\Delta m_j = +1$ or -1	
a)	b)		a)	b)	a)	b)
I'	1'	$\frac{2}{3}\|f_0 - g_0\|^2$	↑	↓	σ^+	σ^-
II'	2'	0				
III'	3'	0				
IV'	4'	$\frac{1}{3}\|f_1 - g_1\|^2$	↑	↓	σ^-	σ^+
V'	5'	$\|f_1 - g_1\|^2$	↑	↓	σ^+	σ^-
VI'	6'	0				
c)	d)		c)	d)	c)	d)
I'	1'	$\frac{2}{3}\|f_0\|^2$	↓	↑	σ^+	σ^-
II'	2'	$\frac{2}{3}\|g_0\|^2$	↑	↓	σ^-	σ^+
III'	3'	$\frac{1}{3}\|g_1\|^2$	↓	↑	σ^+	σ^-
IV'	4'	$\frac{1}{3}\|f_1\|^2$	↓	↑	σ^-	σ^+
V'	5'	$\|f_1\|^2$	↓	↑	σ^+	σ^-
VI'	6'	$\|g_1\|^2$	↑	↓	σ^-	σ^+

Fig. 4.6. Transitions to ${}^2P_{3/2}$ with totally polarized collision partners. a) to d) denote the following processes:
a) e↑ + A↑ → A(${}^2P_{3/2}$) b) e↓ + A↓ → A(${}^2P_{3/2}$)
c) e↓ + A↑ → A(${}^2P_{3/2}$) d) e↑ + A↓ → A(${}^2P_{3/2}$)

take into account the fine-structure splitting since polarization experiments with electron spectrometers of the necessary resolution are not feasible today. Since half the collisions take place with atoms A↑ and the other half with A↓, (factor $\frac{1}{2}$) we obtain, from the relevant processes in (4.28) and (4.30), the cross section for the appearance of e↑

$$\sigma_e^\uparrow({}^2P) = P_e(\tfrac{1}{2}|f_0|^2 + |f_1|^2 + \tfrac{1}{2}|f_0 - g_0|^2 + |f_1 - g_1|^2)$$
$$+ \tfrac{1}{2}(1 - P_e)\sigma({}^2P) \tag{4.33}$$

4.3 Polarization Effects in Inelastic Exchange Scattering   105

with $\sigma(^2P)$ from (4.31). Here the partially polarized primary beam has again been split into a totally polarized and an unpolarized part as in the derivation of (4.22). Similarly, it follows from the lower process in (4.29) that the cross section for appearance of e↓ is

$$\sigma_e^{\downarrow}(^2P) = P_e(\tfrac{1}{2}|g_0|^2 + |g_1|^2) + \tfrac{1}{2}(1 - P_e)\sigma(^2P). \tag{4.34}$$

Thus

$$P_e' = \frac{\sigma_e^{\uparrow} - \sigma_e^{\downarrow}}{\sigma_e^{\uparrow} + \sigma_e^{\downarrow}} = P_e \frac{\sigma(^2P) - |g_0|^2 - 2|g_1|^2}{\sigma(^2P)}.$$

Hence, by measuring $P_e'(\theta)$, one obtains the quantity

$$|g_0(\theta)|^2 + 2|g_1(\theta)|^2 = \sigma(^2P)\left(1 - \frac{P_e'}{P_e}\right), \tag{4.35}$$

if the initial polarization P_e and the differential excitation cross section $\sigma(^2P)$ are known.

By observing the electron polarization P_e' after the inelastic scattering of initially unpolarized electrons by polarized atoms with polarization P_A, one obtains

$$|f_0(\theta)|^2 + 2|f_1(\theta)|^2 = \sigma(^2P)\left(1 - \frac{P_e'}{P_A}\right), \tag{4.36}$$

which can be derived in the same way as (4.35).

Further information on the cross sections for excitation of the various substates is obtained by observing the circularly polarized light which is emitted by the atoms. Now the fine-structure splitting has to be taken into account because it is easily resolved by optical spectrometers. We assume the hyperfine interaction to be decoupled experimentally. Since the 10^{-8} s excited-state lifetime is large compared with the relaxation time of spin-orbit coupling, the analysis of the emitted radiation has to be based on the coupled states $|j, m_j\rangle$.

If one observes the light regardless of the angle at which the electrons are scattered in the excitation process, one does not obtain information on the differential cross sections but rather on the integral cross sections

$$|F_0|^2 = 2\pi \int_0^\pi |f_0(\theta)|^2 \sin\theta d\theta \text{ etc.} \tag{4.37}$$

and

$$Q = 2\pi \int_0^\pi \sigma(\theta) \sin\theta d\theta.$$

The emitted light has linearly and circularly polarized components. Their intensity ratio I^π/I^σ for the transitions from the states $m_j = \pm\frac{1}{2}$ is $\frac{1}{2}$ for $^2P_{1/2} \to {}^2S_{1/2}$ and 2 for $^2P_{3/2} \to {}^2S_{1/2}$. This follows from the calculation of the corresponding transition matrix elements (see for example SCHIFF [4.17], Sect. 48).

Let us first focus attention on the transitions from $^2P_{1/2}$. When all excited atoms return to the ground state by light emission, the cross sections for emission of polarized radiation are

$$I^\pi(^2P_{1/2}) = \tfrac{1}{3}Q(^2P_{1/2}) \tag{4.38}$$

$$I^\sigma(^2P_{1/2}) = I^{\sigma+}(^2P_{1/2}) + I^{\sigma-}(^2P_{1/2}) = \tfrac{2}{3}Q(^2P_{1/2}), \tag{4.39}$$

where $Q(^2P_{1/2})$ is the total cross section for excitation of $^2P_{1/2}$. The last column in Fig. 4.5 indicates the sense of circular polarization of the emitted light, considering that in transitions $\Delta m_j = -1$ and $+1$ from the levels $m_j = +\frac{1}{2}$ and $m_j = -\frac{1}{2}$ circularly polarized σ^+ and σ^- light is emitted. (We will not deal here with linearly polarized light as it can also be obtained by excitation with unpolarized electrons. Only circularly polarized light is typical of excitation by polarized electrons [see (4.42)]).

For excitation of unpolarized atoms by totally polarized electrons, that is, for

$$e\uparrow + \begin{array}{c} A\uparrow \\ A\downarrow \end{array} \to A(^2P_{1/2})$$

which is described by processes a) and d) in Fig. 4.5, we see that

$$\begin{aligned} I^{\sigma+} &= \tfrac{1}{9}|F_0 - G_0|^2 + \tfrac{1}{9}|G_0|^2 + \tfrac{2}{9}|F_1|^2 \\ I^{\sigma-} &= \tfrac{2}{9}|F_1 - G_1|^2 + \tfrac{1}{9}|F_0|^2 + \tfrac{2}{9}|G_1|^2 \end{aligned} \tag{4.40}$$

(where the factors $\tfrac{2}{3}$ from (4.39) and $\tfrac{1}{2}$ due to the scattering on an unpolarized target have been included). Thus for excitation with a partially polarized electron beam of polarization P_e, one has

$$\begin{aligned} I^{\sigma+} &= \tfrac{1}{9}P_e(|F_0 - G_0|^2 + |G_0|^2 + 2|F_1|^2) + \tfrac{1}{2}(1 - P_e)\tfrac{2}{3}Q(^2P_{1/2}) \\ I^{\sigma-} &= \tfrac{1}{9}P_e(2|F_1 - G_1|^2 + |F_0|^2 + 2|G_1|^2) + \tfrac{1}{2}(1 - P_e)\tfrac{2}{3}Q(^2P_{1/2}). \end{aligned}$$
$$\tag{4.41}$$

The circular polarization of the light emitted in the transition

4.3 Polarization Effects in Inelastic Exchange Scattering

$^2P_{1/2} \to {}^2S_{1/2}$ is

$$P^\sigma = \frac{I^{\sigma+} - I^{\sigma-}}{I^{\sigma+} + I^{\sigma-}}$$

$$= P_e \frac{\tfrac{1}{9}(|F_0 - G_0|^2 + |G_0|^2 + 2|F_1|^2 - 2|F_1 - G_1|^2 - |F_0|^2 - 2|G_1|^2)}{\tfrac{2}{3}Q(^2P_{1/2})}. \tag{4.42}$$

Since according to (4.39) and (4.40)

$$Q(^2P_{1/2}) = \tfrac{1}{6}|F_0 - G_0|^2 + \tfrac{1}{3}|F_1 - G_1|^2 + \tfrac{1}{6}|F_0|^2 + \tfrac{1}{3}|F_1|^2 \\
+ \tfrac{1}{6}|G_0|^2 + \tfrac{1}{3}|G_1|^2 \tag{4.43}$$

it follows that

$$P^\sigma Q(^2P_{1/2}) = P_e[Q(^2P_{1/2}) - \tfrac{1}{3}|F_0|^2 - \tfrac{2}{3}|G_1|^2 - \tfrac{2}{3}|F_1 - G_1|^2] \tag{4.44}$$

or

$$\tfrac{1}{3}|F_0|^2 + \tfrac{2}{3}|G_1|^2 + \tfrac{2}{3}|F_1 - G_1|^2 = Q(^2P_{1/2})\left(1 - \frac{P^\sigma}{P_e}\right). \tag{4.45}$$

If we know the polarization P_e of the incident electrons, and the cross section $Q(^2P_{1/2})$ which is found by measuring the total light emitted in the transitions $^2P_{1/2} \to {}^2S_{1/2}$, then, by measuring the circular polarization of the emitted line, we can obtain the sum of three of the unknown cross sections. Thus the sum of the remaining terms in Q is also known: From (4.43) and (4.44) one has

$$P^\sigma Q(^2P_{1/2}) = P_e[\tfrac{2}{3}|F_1|^2 + \tfrac{1}{3}|G_0|^2 + \tfrac{1}{3}|F_0 - G_0|^2 - Q(^2P_{1/2})]$$

or

$$\tfrac{2}{3}|F_1|^2 + \tfrac{1}{3}|G_0|^2 + \tfrac{1}{3}|F_0 - G_0|^2 = Q(^2P_{1/2})\left(1 + \frac{P^\sigma}{P_e}\right). \tag{4.46}$$

Further relations between the cross sections can be obtained in a similar way from Fig. 4.6 by considering the circular polarization of the light emitted in the transitions from $^2P_{3/2}$ (see Problem 4.3).

In order to determine the various unknown cross sections experimentally, other independent combinations of these quantities are needed. Such relations can be obtained by considering the excitation at the threshold energy. The scattered electrons then leave with vanishing energy and hence vanishing orbital angular momentum. Taking the quantization axis parallel to the incident electron beam, we can state that the incident electrons also have vanishing orbital angular momentum along the quantization axis. Hence the electrons neither transfer nor carry away angular momentum parallel to the quantization axis, so that $\Delta m_l = 0$. Consequently, at the threshold energy E_{th}, the transition from the ground state of the alkalis ($l = 0$, $m_l = 0$) can lead only to states with $m_l = 0$. This is different at incident energies well above the threshold energy. In this case the scattered electrons leave with considerable energy and thus with considerable orbital angular momentum. Unless the electrons are scattered at 0° or 180°, they do have orbital angular momentum components along the direction of quantization. This means that usually there is a transfer of orbital angular momentum parallel to the quantization axis at higher energies.

Accordingly, at threshold energy all the cross sections for $m_l = \pm 1$ vanish, whereas with increasing energy the cross sections for $m_l = 0$ tend to zero. For example, we obtain from (4.45) and (4.46)

$$\frac{1}{3}|F_0|^2 = \lim_{E \to E_{th}} Q(^2P_{1/2})\left(1 - \frac{P^\sigma}{P_e}\right)$$
$$\frac{2}{3}|F_1|^2 = \lim_{E \to \infty} Q(^2P_{1/2})\left(1 + \frac{P^\sigma}{P_e}\right)$$
(4.47)

for the $^2P_{1/2} \to {}^2S_{1/2}$ transition.

Further possibilities for determining the cross sections arise from measurements where both collision partners are polarized. Such experiments are very difficult, but have the advantage that the number of possible transitions is further reduced so that the various terms can be more easily separated. In addition, one could observe the scattered electrons in coincidence with the emitted light and thereby determine the polarization of the electrons and of the σ light. These investigations are not within the reach of present experimental techniques; but in principle the individual cross sections can all be determined [4.18] by combining the various possibilities whose results can be read from Figs. 4.5 and 4.6.

Problem 4.3: Equation (4.45) gives the combination of cross sections that is obtained by measuring the circular light polarization in the transition $^2P_{1/2} \to {}^2S_{1/2}$ if unpolarized atoms are excited by polarized electrons. Find the corresponding expression obtained by measuring the circular polarization in the transition $^2P_{3/2} \to {}^2S_{1/2}$.

Solution: In this process, transitions from $m_j = \pm\frac{3}{2}$ result in circularly polarized light only, whereas in transitions from $m_j = \pm\frac{1}{2}$ due to the relation $I^\pi/I^\sigma = 2$ but a third of the emitted light is circularly polarized. Thus one obtains, for totally polarized

4.3 Polarization Effects in Inelastic Exchange Scattering   109

electrons e↑, from processes a) and d) of Fig. 4.6

$$I^{\sigma+} = \frac{1}{9}|F_0 - G_0|^2 + \frac{1}{2}|F_1 - G_1|^2 + \frac{1}{9}|G_0|^2 + \frac{1}{18}|F_1|^2 + \frac{1}{2}|G_1|^2$$

$$I^{\sigma-} = \frac{1}{18}|F_1 - G_1|^2 + \frac{1}{9}|F_0|^2 + \frac{1}{18}|G_1|^2 + \frac{1}{2}|F_1|^2.$$

Hence for partially polarized electrons it follows that

$$I^{\sigma+} = P_e\left(\frac{1}{9}|F_0 - G_0|^2 + \frac{1}{2}|F_1 - G_1|^2 + \frac{1}{9}|G_0|^2 + \frac{1}{18}|F_1|^2 + \frac{1}{2}|G_1|^2\right)$$

$$+ \frac{1}{2}(1 - P_e)I^\sigma$$

$$I^{\sigma-} = P_e\left(\frac{1}{18}|F_1 - G_1|^2 + \frac{1}{9}|F_0|^2 + \frac{1}{18}|G_1|^2 + \frac{1}{2}|F_1|^2\right) + \frac{1}{2}(1 - P_e)I^\sigma,$$

where

$$I^\sigma = I^{\sigma+} + I^{\sigma-} = \frac{1}{9}|F_0 - G_0|^2 + \frac{5}{9}|F_1 - G_1|^2 + \frac{1}{9}|F_0|^2 + \frac{5}{9}|F_1|^2$$

$$+ \frac{1}{9}|G_0|^2 + \frac{5}{9}|G_1|^2$$

is independent of the polarization of the incident electrons. This yields

$$P^\sigma = \frac{I^{\sigma+} - I^{\sigma-}}{I^{\sigma+} + I^{\sigma-}}$$

$$= \frac{P_e}{9I^\sigma}(|F_0 - G_0|^2 + 4|F_1 - G_1|^2 + |G_0|^2 + 4|G_1|^2 - 4|F_1|^2 - |F_0|^2). \quad (4.48)$$

Contrary to what might be expected from (4.42), $I^\sigma(^2P_{3/2})$ and $Q(^2P_{3/2})$ are not connected by a fixed numerical factor. Reason: the fraction of excitations which lead to $m_j = \pm 3/2$ (σ light only) and to $m_j = \pm 1/2$ (fraction of σ light is 1/3) is not specified by $Q(^2P_{3/2})$. It depends on the size of the individual terms of $Q(^2P_{3/2})$ which thus also determine the fraction of circularly polarized light. From (4.48) it follows that

$$P^\sigma I^\sigma = P_e\left(I^\sigma - \frac{1}{9}|F_1 - G_1|^2 - \frac{2}{9}|F_0|^2 - |F_1|^2 - \frac{1}{9}|G_1|^2\right) \quad \text{or}$$

$$\frac{1}{9}|F_1 - G_1|^2 + \frac{2}{9}|F_0|^2 + |F_1|^2 + \frac{1}{9}|G_1|^2 = I^\sigma\left(1 - \frac{P^\sigma}{P_e}\right).$$

To determine the left-hand side of the last equation one needs to measure P^σ, P_e, and the total cross section for producing circularly polarized light by excitation of the $^2P_{3/2}$ state.

4.3.2 Two-Electron Atoms

Experiments with polarized atoms, such as those discussed in the preceding section, cannot be made with atoms that have two outer electrons and saturated spins (spin quantum number $S = 0$). These atoms are in singlet states which cannot be polarized. We shall see, however, that experiments with polarized electrons still give valuable insights into collision processes with such atoms.

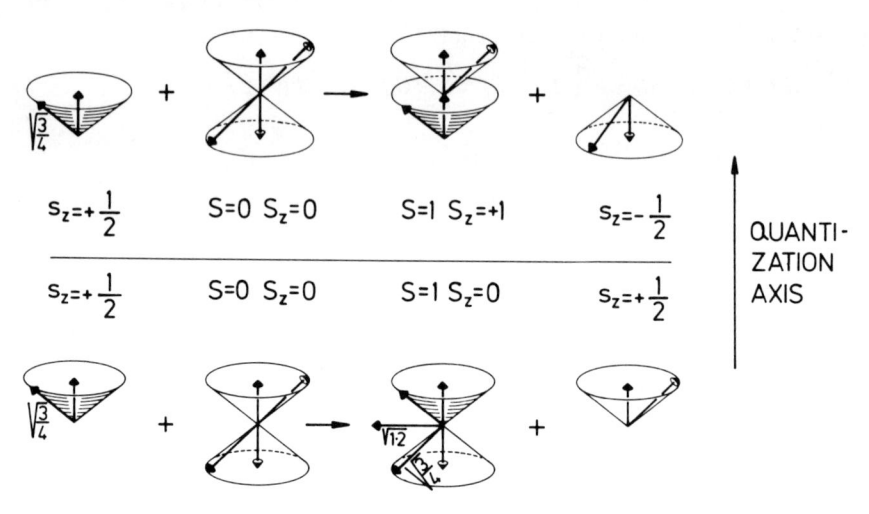

4. Exchange Processes in Electron-Atom Scattering

Fig. 4.7. Excitation of triplet states by exchange collisions

Let us consider the excitation of a triplet state ($S = 1$) of such a two-electron atom by polarized electrons e↑. As before, we will first exclude forces that are explicitly spin-dependent. This is a good approximation for light atoms like helium. Then the excitation of a triplet state from the singlet ground state can occur only by exchange of the incident electron with one of the atomic electrons as shown in Fig. 4.7. We see from the figure that in this process the orientation quantum numbers have the following values:

	m_s(electron)	M_S(atom)
before the collision	$+\frac{1}{2}$	0
after the collision	either $+\frac{1}{2}$	0
	or $-\frac{1}{2}$	1

The polarization direction of the electrons has been chosen as the reference axis. We see that with e↑ the state $M_S = -1$ cannot be reached without violating the conservation of spin angular momentum.

It follows from what was shown in the preceding section that one has a circular polarization of the emitted light: In diagrams analogous to Figs. 4.5 and 4.6, where the substates M_J are expressed in terms of M_S and M_L ($M_S + M_L = M_J$), the states $M_S = -1$ are not populated by excitation with e↑. Since these states are located preferentially on the left-hand side of the diagrams one has a significant disparity between the populations of

4.3 Polarization Effects in Inelastic Exchange Scattering

the substates with positive and negative M_J. This results in a preferential emission of σ^+ light. For e↓ the situation is reversed so that mainly σ^- light is emitted. The circular light polarization can be utilized to find the cross sections for excitation of the sublevels as described in the preceding section. It can, of course, also be used to determine the spin polarization of the exciting electrons [4.19, 20], once the quantitative relations between electron polarization and circular light polarization have been established experimentally. This is one of the numerous examples in the field of polarized electrons where simple facts are set down in the literature without the much more difficult task of experimental verification having been achieved.

Let us now consider the polarization of the scattered electrons. The cross section for excitation of $M_S = 1$ is twice as large as that for excitation of $M_S = 0$. This can be seen from a simple calculation (see Problem 4.4) and is made plausible by Fig. 4.8, which represents the excitation by an unpolarized electron beam. The e↑ of this beam excite the levels $M_S = +1, 0$, whereas the e↓ excite $M_S = -1, 0$. Since $M_S = 0$ can be excited by the e↑ of the unpolarized beam as well as by the e↓, one would obtain a disparity in the populations of $M_S = 0$ and $M_S = \pm 1$ if the excitation cross sections for $M_S = +1$ and -1 were not twice as large as that for excitation of $M_S = 0$. There is no reason why such an alignment of the spin directions should occur by excitation with an unpolarized beam. We have seen in Fig. 4.7 that in excitation of a state $M_S = +1$ there is a reversal of the free-electron spin direction, whereas in excitation of $M_S = 0$ the spin directions of incident and scattered electrons are the same. Consequently, excitation by a totally polarized electron beam e↑ yields a scattered beam with two thirds e↓ and one third e↑, in other words, a polarization $P' = -\frac{1}{3}$ of the scattered beam.

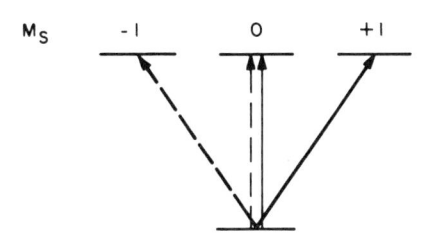

Fig. 4.8. Excitation of sublevels $M_S = 0, \pm 1$ (which may be part of sublevels M_J like in Figs. 4.5 and 4.6) by unpolarized electrons. Solid lines: excitation by e↑; broken lines: excitation by e↓

Since the derivation of this result was straightforward, it does not seem very challenging to make such an experiment because one knows its result in advance. For helium, for example, where the underlying assumption (no explicit spin-dependent forces) is well established, it seems quite obvious that a measurement of the ratio of the polarizations P' of the scattered beam and P of the incident beam would yield $P'/P = -\frac{1}{3}$. This assumes, of course, that there is no other influence on the polarization which has not been considered here (e.g., by compound states). The first polarization experiment of this kind has therefore been made with a target with which there was no question that one would learn something new:

For mercury it is well known that spin-dependent forces (spin-orbit forces) are no longer negligible. The excitation of a triplet state can in this case occur not only by an exchange of electrons but also by a direct process, in which the spin of one of the atomic electrons flips during excitation. This affects, of course, the value of P'/P so that a measurement of this ratio yields the extent to which the exchange processes discussed above still contribute to the excitation. This has been studied in a triple scattering experiment [4.21, 22].

Fig. 4.9. Triple scattering experiment for direct observation of exchange excitation in mercury [4.21]

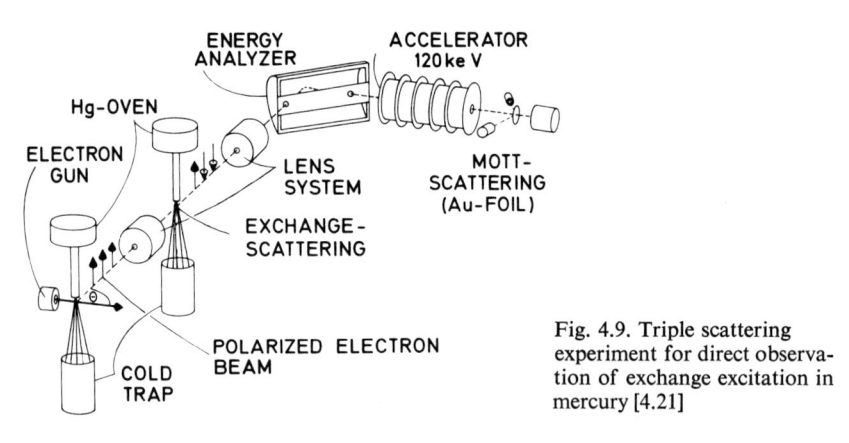

Figure 4.9 is a schematic diagram of the apparatus. Scattering from a mercury-vapor beam, as described in Subsection 3.6.1, is used to produce a polarized electron beam of 80 eV and $P = 0.22$. The polarized electrons are decelerated to energies between 5 and 15 eV and focused on a second mercury target. From the electrons scattered here, an energy analyzer selects those that have been scattered in the forward direction after excitation of the $6\,^3P$ states of the mercury atoms (energy loss ~ 5 eV). The polarization of these electrons is measured by a Mott detector.

There were two reasons for studying the electron scattering in the forward direction. First, there is the maximum of the scattered intensity; this is essential for a triple scattering experiment with its notorious lack of intensity. Second, the spins of the electrons scattered in the forward direction are not affected by spin-orbit coupling (see Sect. 3.5 where this has been illustrated by the fact that electrons scattered at small angles pass by the atom at a distance large enough that spin-orbit interaction is negligible). It was quite uncertain whether P' for electrons scattered in the forward direction would be different from the incident polarization P, because the theoretical treatment of exchange scattering in the forward direction was particularly difficult [4.23, 24] and yielded no reliable results. Since it is the exchange processes that cause the change of the free-electron spin direction, P' would equal P if these processes did not play a role at small scattering angles.

The experimental results of Fig. 4.10 show that at incident energies below 8 eV there is a great number of processes which change the polarization, and at 6 eV the limiting value of $P'/P = -\frac{1}{3}$ is even observed (within the experimental error limits). That means that at this energy nearly all the excitation processes of the 6 ^3P levels occur by exchange scattering. On the other hand, the exchange excitation discussed above no longer plays an appreciable role at energies above 10 eV. Figure 4.11 gives these facts directly for the 6 $^3P_1$ state. It is the evaluation of a measurement in which the fine structure of 6 ^3P has been resolved.

Needless to say, experiments of this kind are rather delicate and need careful checks in order to ensure that the observed depolarization is not spurious. An essential check of the experiment discussed is shown in Fig. 4.12. Here the excitation of the 6 $^1P_1$ level (energy loss 6.7 eV) has been

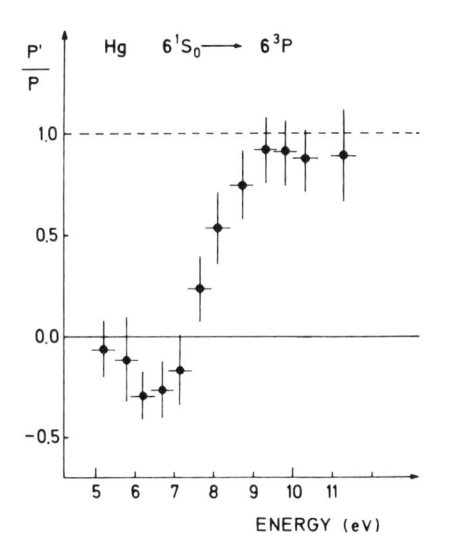

Fig. 4.10. Measured values [4.21] of depolarization vs. incident energy for 6 $^1S_0 \rightarrow 6$ ^3P (forward direction)

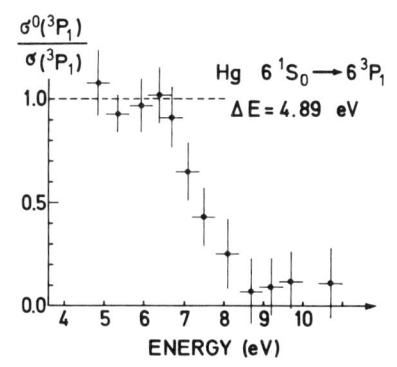

Fig. 4.11. Contribution of the exchange processes discussed above to the excitation of the 6 $^3P_1$ state [4.21]. σ^0 is the differential cross section for excitation by these exchange processes, σ is the complete differential cross section for excitation of 6 $^3P_1$. All results in forward direction

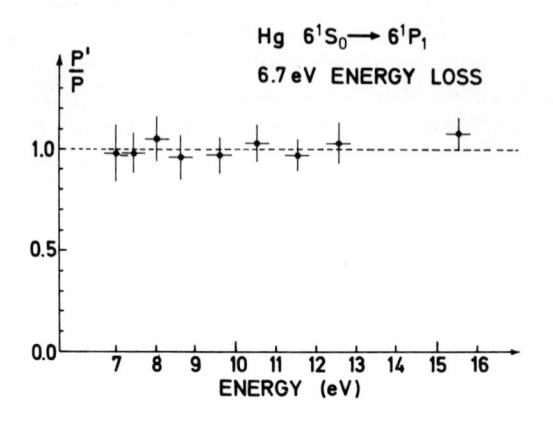

Fig. 4.12. Measured values [4.21] of depolarization vs. incident energy for $6\,^1S_0 \to 6\,^1P_1$ (forward direction)

studied in the same apparatus. In the excitation of a singlet state from a singlet ground state no change of spin directions can occur, no matter whether the excitation takes place by a direct or an exchange process.[5] (Remember that we are discussing cases where polarization effects due to spin-orbit interaction of the unbound electrons are negligible). This is observed in the experiment, which shows that no spurious depolarization effects occur.

The derivation of the formulae expressing the polarization of the final beam in terms of the direct and exchange scattering amplitudes may be performed by the same method discussed in the preceding subsection and will therefore not be given here. It can also be found in the original paper [4.22], where density-matrix techniques have been used. It should be pointed out, however, that the assumption "excitation time ≪ spin-orbit relaxation time" made in Subsection 4.3.1 is no longer valid when one resolves the fine structure caused by the spin-orbit coupling of the atom. In this case the excitation leads immediately to the spin-orbit coupled states.

We have seen in this chapter that experiments with polarized electrons yield direct information on exchange scattering, a field about which quantitative information is meager. The method takes advantage of the fact that electrons which are labeled by different spin directions are distinguishable, if they retain their spin direction during the scattering event to be studied. General theoretical results are now available, even for cases that are beyond the reach of present experimental techniques [4.25]. Calculations and experiments that put these general relations to quantitative test are extremely scarce and highly desirable.

---

[5] The same is true, of course, for elastic scattering from atoms in a singlet ground state; there is no point in studying elastic exchange scattering from such atoms with polarized electrons.

4.3 Polarization Effects in Inelastic Exchange Scattering   115

The discussion of the present chapter has been restricted to those cases in which such quantitative information is either available, though limited, or can be expected in the not too distant future. Further relevant work on exchange scattering with polarized electrons will be discussed in Section 7.3 in connection with sources of polarized electrons.

Problem 4.4: Show that the cross section for excitation of a triplet state with $M_S = 1$ from a singlet ground state is twice as large as that for excitation of $M_S = 0$, provided that the other quantum numbers of these states are identical.

Solution: We evaluate the scattering amplitude (4.4) with the antisymmetric wave functions

$$\psi_i = \frac{1}{\sqrt{3}}[\exp(i\mathbf{k}\mathbf{r}_1)\eta(1)u(\mathbf{r}_2,\mathbf{r}_3)\chi_A(2,3) + \exp(i\mathbf{k}\mathbf{r}_2)\eta(2)u(\mathbf{r}_3,\mathbf{r}_1)\chi_A(3,1)$$
$$+ \exp(i\mathbf{k}\mathbf{r}_3)\eta(3)u(\mathbf{r}_1,\mathbf{r}_2)\chi_A(1,2)]$$

and

$$\psi_f = \frac{1}{\sqrt{3}}[\exp(i\mathbf{k}'\mathbf{r}_1)\eta'(1)u'(\mathbf{r}_2,\mathbf{r}_3)\chi_S(2,3) + \exp(i\mathbf{k}'\mathbf{r}_2)\eta'(2)u'(\mathbf{r}_3,\mathbf{r}_1)\chi_S(3,1)$$
$$+ \exp(i\mathbf{k}'\mathbf{r}_3)\eta'(3)u'(\mathbf{r}_1,\mathbf{r}_2)\chi_S(1,2)],$$

where the notation is the same as in Section 4.1; χ_S is given by (4.15a) and (4.15b) for the final substates $M_S = +1$ and 0, respectively. Multiplication yields nine terms. The three terms corresponding to direct scattering (incoming and scattered electrons have the same label) disappear because $\chi_A(\lambda,\mu)$ is orthogonal to $\chi_S(\lambda,\mu)$. If we first consider the transitions to $M_S = +1$, the six non-vanishing terms have the form

$$\frac{g(\theta)}{3\sqrt{2}}\{\eta'(2)\alpha(3)\alpha(1)\eta(1)[\alpha(2)\beta(3) - \beta(2)\alpha(3)]\},$$

where

$$g(\theta) = -\frac{m}{2\pi\hbar^2}\langle\exp(i\mathbf{k}'\mathbf{r}_2)u'(\mathbf{r}_3,\mathbf{r}_1)|T|\exp(i\mathbf{k}\mathbf{r}_1)u(\mathbf{r}_2,\mathbf{r}_3)\rangle, \qquad (4.49)$$

etc., through permutation.[6] The spin function of the incident e↑ is $\eta = \alpha$. If the spin function η' of the outgoing electrons also equals α, the above product of the spin functions is zero. This is in agreement with Fig. 4.7 which shows a change of the spin direction of the free electron for excitation of $M_S = +1$. For $\eta' = \beta$ the product of the spin functions is -1. Since we have six terms of the kind given above, we find the scattering amplitude to be $-6[g(\theta)/3\sqrt{2}] = -\sqrt{2}g(\theta)$.

For $M_S = 0$ the six non-vanishing terms have the form

$$\frac{g(\theta)}{3\cdot 2}\{\eta'(2)[\alpha(3)\beta(1) + \beta(3)\alpha(1)]\alpha(1)[\alpha(2)\beta(3) - \beta(2)\alpha(3)]\},$$

etc., through permutation. This time the spin product is zero for $\eta' = \beta$, in accordance with Fig. 4.7 which shows no change of the free-electron spin for excitation of $M_S = 0$. For $\eta' = \alpha$ the spin product is 1, so that the sum of the six non-vanishing terms yields the scattering amplitude $g(\theta)$.

The expressions (4.49) for $g(\theta)$ are identical for excitation of $M_S = 1$ and $M_S = 0$ if the space functions of the final states are identical. The ratio of the scattering amplitudes is therefore $-\sqrt{2}g(\theta)/g(\theta) = -\sqrt{2}$, and the ratio of the cross sections is 2.

---

[6] In permuting the electrons one has to take into account that the u are symmetric and the u' are antisymmetric.

4.4 Møller Scattering

Electron-electron scattering is spin dependent. This can be utilized for polarization analysis. Comparison with the Mott analyzer is made.

One of the few cases where reliable information on polarization effects in inelastic scattering is available is electron scattering by electrons whose binding energy is small compared to the energy transfer during the collision. One can then neglect the binding energy and consider the scattering process as an elastic collision between free electrons. We can therefore use (4.7) for the scattering amplitudes and replace the wave functions u and u' by free-electron wave functions, since bonding is no longer considered. Consequently, we obtain in the first Born approximation

$$f = -\frac{m}{2\pi\hbar^2}\left\langle \exp(i\mathbf{k}'_1\mathbf{r}_1)\exp(i\mathbf{k}'_2\mathbf{r}_2)\left|\frac{e^2}{r_{12}}\right|\exp(i\mathbf{k}_1\mathbf{r}_1)\exp(i\mathbf{k}_2\mathbf{r}_2)\right\rangle$$
$$g = -\frac{m}{2\pi\hbar^2}\left\langle \exp(i\mathbf{k}'_1\mathbf{r}_2)\exp(i\mathbf{k}'_2\mathbf{r}_1)\left|\frac{e^2}{r_{12}}\right|\exp(i\mathbf{k}_1\mathbf{r}_1)\exp(i\mathbf{k}_2\mathbf{r}_2)\right\rangle,$$
(4.50)

where the scattering potential e^2/r_{12} has been introduced. In the center-of-mass system we have $\mathbf{k}_1 = -\mathbf{k}_2$, $\mathbf{k}'_1 = -\mathbf{k}'_2$ so that, after separation of the motion of the center of mass, we obtain

$$f = -\frac{m}{2\pi\hbar^2}\int \exp[i(\mathbf{k}_1 - \mathbf{k}'_1)\mathbf{r}]\frac{e^2}{r}d^3r$$
$$g = -\frac{m}{2\pi\hbar^2}\int \exp[i(\mathbf{k}_1 + \mathbf{k}'_1)\mathbf{r}]\frac{e^2}{r}d^3r,$$
(4.51)

where $\mathbf{r} = \mathbf{r}_1 - \mathbf{r}_2$. Since $|\mathbf{k}_1 - \mathbf{k}'_1| = 2k_1 \sin\theta_c/2$, $|\mathbf{k}_1 + \mathbf{k}'_1| = 2k_1 \cos\theta_c/2$, where θ_c is the scattering angle in the center-of-mass system, one easily sees, by taking the polar axis in the direction of the vectors $\mathbf{k}_1 - \mathbf{k}'_1$ and $\mathbf{k}_1 + \mathbf{k}'_1$, respectively, that

$$g(\theta_c) = f(\pi - \theta_c).$$

Let us assume that a totally polarized electron beam is scattered by a totally polarized target, the angle between the polarization of the target and that of the incident beam being ϑ. Then we find from (4.26) the cross section

$$\sigma(\theta_c) = \tfrac{1}{2}[(1 + \cos \vartheta)|f(\theta_c) - f(\pi - \theta_c)|^2$$
$$+ (1 - \cos \vartheta)(|f(\theta_c)|^2 + |f(\pi - \theta_c)|^2)]. \quad (4.52)$$

For the scattering angle $\theta_c = 90°$ we obtain

$$\sigma(90°) = (1 - \cos \vartheta)|f(90°)|^2,$$

which is 0 for parallel spins ($\vartheta = 0$) and $2|f(90°)|^2$ for antiparallel spins ($\vartheta = 180°$).

If the target electron has low binding energy, it can be considered to be at rest in the laboratory system. Since θ_c and the scattering angle θ in the laboratory system are related by $\theta_c = 2\theta$ we have $\sigma_p/\sigma_a = 0$ for $\theta = 45°$, when σ_a and σ_p are the differential cross sections for parallel and antiparallel spins. This case where half of the incident energy is transferred to the target electron (see Fig. 4.13) is therefore most suitable for polarization analysis: polarized electrons may be scattered by a magnetic material whose spins are oriented first parallel and then antiparallel to the incident polarization. The relative difference of the scattering intensity (asymmetry) yields the unknown polarization.

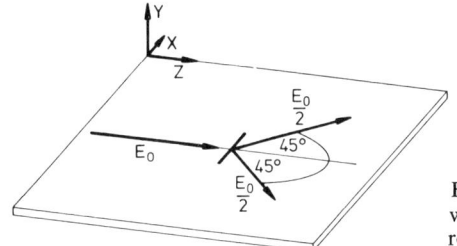

Fig. 4.13. Electron-electron scattering with symmetric energy transfer (non-relativistic limit)

A relativistic treatment of electron-electron scattering does not, in principle, affect this scattering asymmetry which is caused by electron exchange; the numerical results are, however, modified. Only at low velocities is the interaction of the electrons, at each instant of time, given by their static interaction. At higher electron velocities there are retardation effects of the electromagnetic interaction due to the finite velocity of light. Exchange of photons between the two electrons then becomes important, so that one has no longer a problem of quantum mechanics but, instead, of quantum electrodynamics.

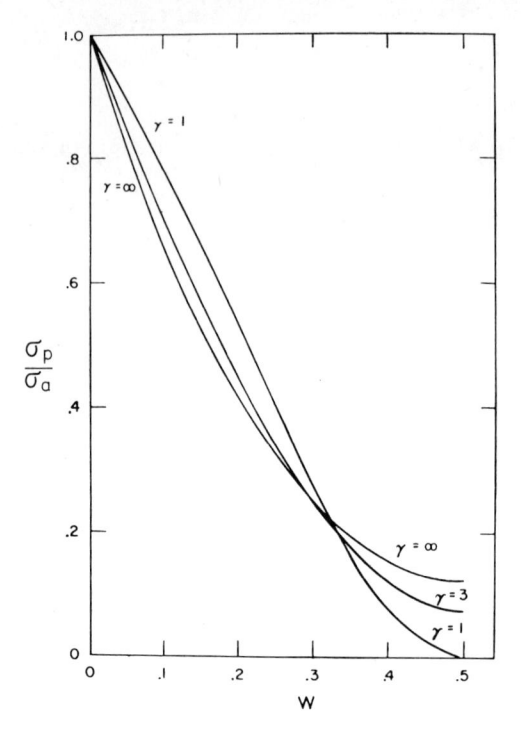

Fig. 4.14. Ratio of the differential cross sections for scattering of longitudinally polarized electrons with parallel (σ_p) and antiparallel (σ_a) spins [4.26]

A solution to this problem under the aspect considered here has been given by BINCER [4.26] who calculated the spin dependence of the cross section for scattering of two Dirac electrons. The first treatment of relativistic electron-electron scattering by MØLLER [4.21] did not emphasize polarization phenomena. Bincer's results for longitudinally polarized electrons are shown in Fig. 4.14, where σ_p/σ_a is given as a function of the energy transfer $w = W/T$ (W is the kinetic energy lost in the collision by the incident electron and $T = mc^2(\gamma - 1)$ is its kinetic energy before the collision). As explained above, one has in the non-relativistic limit $\sigma_p/\sigma_a = 0$ at $w = 0.5$. For higher energies σ_p/σ_a still has a minimum at $w = 0.5$, though its value increases from 0 to 1/8 in the extreme relativistic limit ($\gamma \to \infty$). The scattering angle θ belonging to $w = 0.5$ decreases from 45° in the non-relativistic limit $\gamma = 1$ to 0° for $\gamma \to \infty$.

In our discussion of the spin dependence of the cross section in the non-relativistic case no assumption has been made about the direction of the spin relative to the momentum. Accordingly, the results are valid both for longitudinal and for transverse polarization. This becomes different in the relativistic region, as Fig. 4.15 illustrates. The meaning of the asymmetry

4.4 Møller Scattering

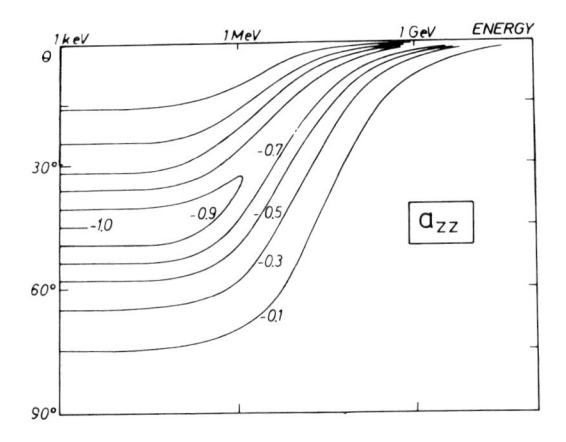

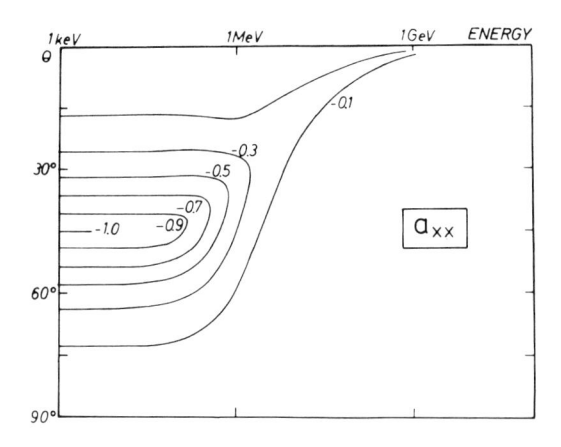

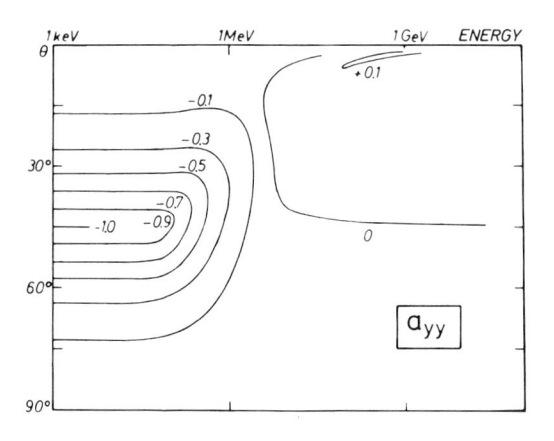

Fig. 4.15. Contours of the asymmetry coefficients for electron-electron scattering (cf. Eq. (4.53)) [4.28]

coefficients a_{ij} shown there can be seen if one writes the cross section in the form

$$\sigma(\theta) = I(\theta)(1 + a_{ij})P_i P_j^{(t)} \tag{4.53}$$

where $I(\theta)$ is the Møller cross section for unpolarized electrons and P_i and $P_j^{(t)}$ ($i, j = x, y, z$) are the polarization components of the incident electron and the target electron, respectively. The direction of the axes is as shown in Fig. 4.13.

The contours a_{zz} = const. again illustrate the spin dependence of the cross section for scattering of longitudinally polarized electrons. In the non-relativistic region, the spin dependence for transversely polarized electrons is the same as can be seen from the contours a_{xx} = const and a_{yy} = const. Between 100 keV and 1 MeV there is, however, a strong decrease in a_{xx} and a_{yy}, so that above 1 MeV the spin dependence of scattering for transversely polarized electrons is considerably lower than that for longitudinally polarized electrons.

The coefficients a_{ij} ($i \ne j$) which describe scattering of transversely polarized by longitudinally polarized electrons or vice versa are, at all energies and angles, either smaller than 0.1 or vanish. This means that in those cases, the spin dependence of scattering is negligible, particularly if one considers the experimental conditions under which these effects can be studied: even in an iron target that is magnetized to saturation, only approximately 8% ($\approx 2/26$) of the electrons are orientated. A target polarization $P^{(t)} = \pm 8\%$ yields, with the optimum value of $a_{ii} = -1$ in (4.53), a scattering asymmetry of only 8%. With asymmetry coefficients that are much smaller, the observed effects are reduced below the 1% limit, which makes experiments cumbersome.

The smallness of the observable effects is the reason why Møller scattering has not become the predominant method for measuring electron polarization although it certainly has several advantages. Many studies of spin polarization in β decay utilized this technique, taking advantage of the fact that the longitudinal polarization of the electrons need not be converted into transverse polarization as in Mott scattering, if the target is magnetized in the direction of the incident spins. The few electron-electron scattering events can be selected from the many background electrons, which are mainly due to scattering from the nuclear Coulomb field[7], by detecting the two outgoing electrons in coincidence. This is done

---

[7] The scattering intensity in electron-electron scattering is proportional to the number Z of electrons per atom, whereas the intensity scattered from the nuclear Coulomb field is, roughly speaking, proportional to Z^2.

by two counters arranged under suitable angles (45° in the non-relativistic range, as shown in Fig. 4.13; smaller angles at higher energies). Pulse-height analysis further distinguishes the two electrons sharing the incident energy from the electrons scattered by the Coulomb field. The technique described also reduces the influence of multiple scattering: electrons that undergo considerable scattering will not be recorded.

The experiments discussed in the various chapters of this book show however that Mott scattering is more widely used than Møller scattering for polarization analysis. The asymmetry effects are larger for the former and careful absolute measurements of the asymmetry function have been made, so that one does not have to rely on theory for the calibration of the polarization analyzer. Although careful measurements of the Møller cross section for unpolarized particles have been made [4.29, 30], absolute measurements of the asymmetry coefficients are not known to the author, so that one has more or less to rely on theoretical results, such as those shown in Fig. 4.15, which are based on lowest order perturbation theory. On the other hand, Møller scattering does not have the disadvantage of Mott scattering that it is inapplicable in the extreme relativistic region. Since $\sigma_p/\sigma_a = 1/8$ even for $\gamma \to \infty$, a polarization analyzer based on this method works at all energies (see Subsect. 7.1.5).

It is quite obvious from our discussion that the employment of Møller scattering would not lead to an efficient source of polarized electrons. Even with an iron target the polarization would never be larger than 8%, and the necessity of discriminating against the large number of background electrons would make such a source even less attractive.

5. Polarized Electrons by Ionization

5.1 Photoionization of Polarized Atoms

Polarized electrons can be produced by photoionization of polarized atomic beams, which have high intensities when produced by six-pole magnets. The process has been investigated mainly with a view to building a source of polarized electrons.

In the first chapter it was shown that a Stern-Gerlach magnet cannot be used as a polarization filter to select free electrons with a certain spin direction. Electrons that are bound to atoms can, however, be polarized in this way. If the oriented atomic electrons are extracted from the atoms without affecting their spin directions, polarized free electrons are obtained. This can, for example, be achieved by photoionization.

Although such an experiment was suggested [5.1] as early as 1930, it has been realized only in recent years [5.2, 3]. Instead of using conventional Stern-Gerlach magnets, six-pole magnets (see Fig. 5.1) were used to polarize alkali atoms. The atoms with the required spin orientation, emerging divergently from the atomic beam oven, can thereby be focused so that high intensities can be attained. The reason for this is as follows:

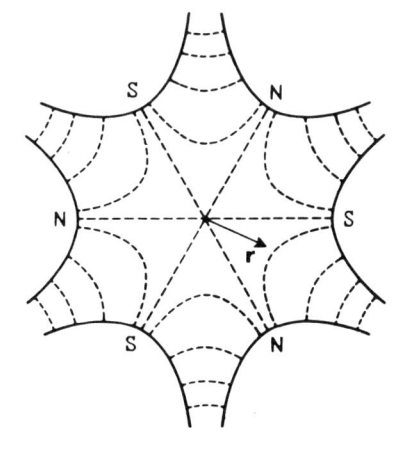

Fig. 5.1. Field of six-pole magnet

Let us assume a field that exerts a force proportional to $\mp\mu r$ on the magnetic dipoles, where $\mp$ refer to directions of the electron spin parallel and antiparallel to the magnetic field, and r is the distance from the axis. Such a field deflects away from the axis those atoms whose spins are antiparallel to its direction; it acts on them as a diverging lens. Atoms with the opposite spin directions are deflected towards the axis. They perform harmonic oscillations of uniform frequency about the axis and are therefore focused to one point if they have equal axial velocity. The field acts on these atoms as a converging lens.

It can be shown that a six-pole magnet as drawn in Fig. 5.1 possesses such lens properties: Near the axis, the magnitude $|B|$ of the magnetic induction is to a good approximation proportional to r^2 (B itself is of course in no way axially symmetric, as one can see from Fig. 5.1). Thus the potential energy of the dipoles for the two spin directions is

$$V = \pm\mu|B| \propto \pm\mu r^2,$$

and the force is $-\nabla V \propto \mp\mu r$. In actual fact, the magnetic field and thus also the spins parallel to it have all possible directions. If, however, a magnetic field in the direction of the axis is attached to the six-pole magnet, the field as well as the spins lying parallel to the field gradually turn into the axial direction as one goes from the inside of the magnet to the outside. The change of the magnetic field direction as seen from the moving particles takes place slowly in comparison to the Larmor frequency so that the spins follow the change of the magnetic field adiabatically.

In this way the six-pole magnet produces a longitudinally polarized, well-focused beam of alkali atoms. The oriented valence electrons are then ejected by photoionization. This occurs within the axial magnetic field just mentioned which is made strong enough to decouple the atomic electron spin s from the nuclear spin I ($j = s = \frac{1}{2}$ in the ground state of alkali atoms). Without decoupling, the resultant angular momentum $F = s + I$ and not the electron spin would be oriented in the magnetic field.[1] The spin expectation value in the direction of orientation of the selected atoms would then decrease; in other words, the observed polarization would be diminished.

After extraction of the photoelectrons from the region of the magnetic decoupling field they were, in the aforementioned experiments, sent through a polarization transformer in order to convert their longitudinal polarization into transverse polarization, as required for analysis by a

---

[1] This is analogous to the coupling of s and l to j which will be discussed in more detail in Section 5.2.

Mott detector. The maximum polarization obtained was close to 80%. The experiments were not done primarily because of an interest in the underlying physical processes, but rather for the purpose of building an intense source of polarized electrons. A further discussion of this method will therefore be given in Section 7.3, so that we can refrain here from giving further experimental details.

5.2 Fano Effect

Polarized electrons can be produced by photoionization of unpolarized alkali atoms with circularly polarized light. This effect (Fano effect) which arises from the spin-orbit coupling of the unbound states is described theoretically and illustrated. Degrees of polarization up to 100% have been found experimentally.

5.2.1 Theory of the Fano Effect

The obvious idea that polarized electrons could be obtained by photo-ionization of polarized atoms was put forward long ago. That the same goal could be achieved with less effort by starting with unpolarized atoms and using circularly polarized light is more difficult to see and was first recognized by FANO [5.4] in 1969.

We describe the Fano effect with the aid of Fig. 5.2, which gives the relevant energy levels of an alkali atom.[2] The unpolarized atomic beam is a mixture of equal numbers of atoms A↑ and A↓ with spins parallel and antiparallel to the quantization axis, which we assume to be parallel to the direction of light propagation. This means that the levels $m_j = m_s = +\frac{1}{2}$ and $-\frac{1}{2}$ of the ground state $^2S_{1/2}$ are equally populated.

The transitions caused by the incident light lead to P states because of the selection rule $\Delta l = \pm 1$. For alkali atoms, the P states have the total angular momenta $j = \frac{1}{2}, \frac{3}{2}$. Radiation of a wavelength short enough for ionization leads to transitions into the continuous $P_{1/2}$ and $P_{3/2}$ states adjoining the bound $P_{1/2}$ and $P_{3/2}$ states at the ionization threshold. If the unpolarized atomic beam is ionized by circularly polarized σ^+ light one has the additional selection rule $\Delta m_j = +1$. One then obtains the transitions 1, 2, and 3 shown in Fig. 5.2.

The final-state angular momenta, which are our main interest, can be directly seen from the spin and angular parts of the wave functions. For the

---

[2] It has been shown [5.4] that the influence of hyperfine interaction on these polarization effects is negligible.

126   5. Polarized Electrons by Ionization

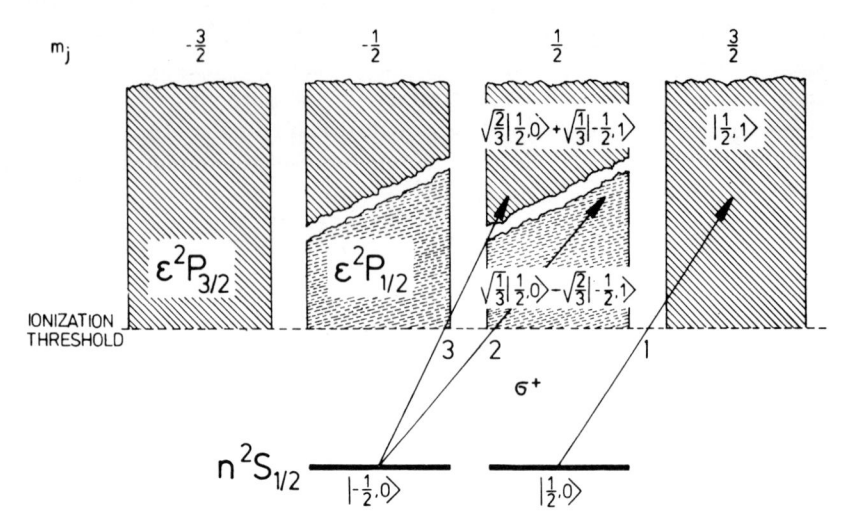

Fig. 5.2. Level diagram for the discussion of the photoionization of alkali atoms. The angular-momentum properties of the states are characterized by combinations of the kets $|m_s, m_l\rangle$

final state reached via transition 1 this part is

$$\begin{pmatrix} 1 \\ 0 \end{pmatrix} Y_{1,1}(\theta, \phi) \tag{5.1}$$

(see Problem 5.1), if $Y_{lm}(\theta, \phi)$ denote the spherical harmonics. From the spin function $\begin{pmatrix} 1 \\ 0 \end{pmatrix}$ it can be seen that the spin component in the z direction (quantization axis) is $+\frac{1}{2}$; the eigenvalues of $Y_{1,1}$ are given by $l = m_l = 1$. We therefore have abbreviated this state in Fig. 5.2 as $|m_s, m_l\rangle = |\frac{1}{2}, 1\rangle$. Its vector-model representation is given in Fig. 5.3a.

The spin directions of the states reached via transitions 2 and 3 can no longer be simply described by a single quantum number. The state $m_j = \frac{1}{2}$, for example, can be realized by $m_s = \frac{1}{2}$, $m_l = 0$ or by $m_s = -\frac{1}{2}$, $m_l = +1$. But neither of the corresponding eigenfunctions of j_z, $\begin{pmatrix} 1 \\ 0 \end{pmatrix} Y_{1,0}(\theta, \phi)$ and $\begin{pmatrix} 0 \\ 1 \end{pmatrix} Y_{1,1}(\theta, \phi)$, is simultaneously an eigenfunction of j^2, which, owing to the conservation of total angular momentum, is required for a realistic wave function. Using the Clebsch-Gordan coefficients, the correct eigenfunctions can, however, be constructed as linear

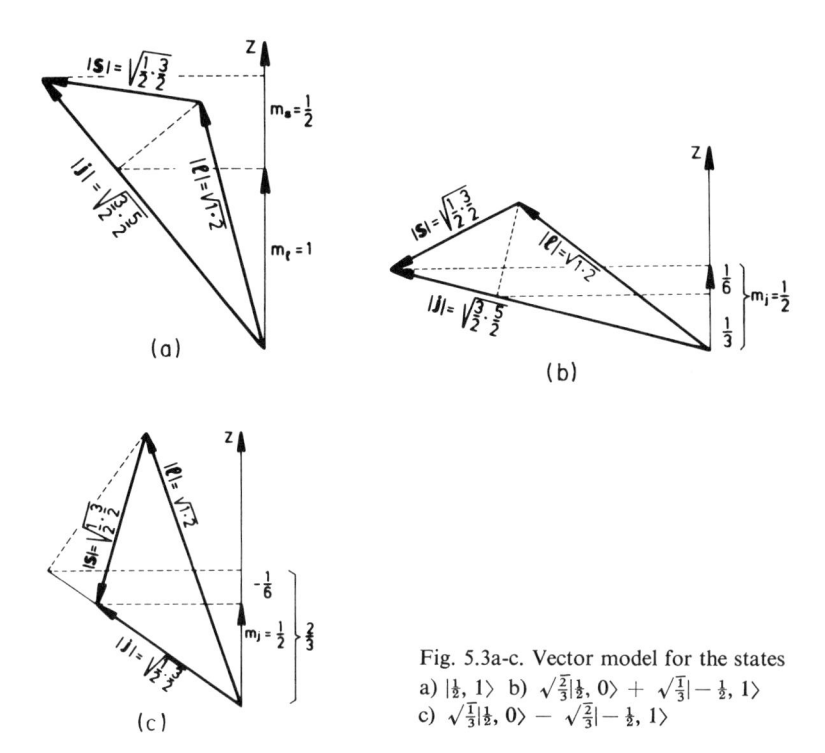

Fig. 5.3a-c. Vector model for the states
a) $|\frac{1}{2}, 1\rangle$ b) $\sqrt{\frac{2}{3}}|\frac{1}{2}, 0\rangle + \sqrt{\frac{1}{3}}|-\frac{1}{2}, 1\rangle$
c) $\sqrt{\frac{1}{3}}|\frac{1}{2}, 0\rangle - \sqrt{\frac{2}{3}}|-\frac{1}{2}, 1\rangle$

combinations of these two parts. For $^2P_{3/2}(m_j = \frac{1}{2})$ one finds

$$\sqrt{\frac{2}{3}}\binom{1}{0}Y_{1,0}(\theta, \phi) + \sqrt{\frac{1}{3}}\binom{0}{1}Y_{1,1}(\theta, \phi). \tag{5.2}$$

The reader who is not familiar with Clebsch-Gordan coefficients can easily check by direct calculation that (5.2) is an eigenfunction of j^2 and j_z with the eigenvalues $j(j + 1) = \frac{3}{2} \cdot \frac{5}{2}$ and $m_j = \frac{1}{2}$ (see Problem 5.1).

According to the quantum mechanical interpretation of the expansion of a wave function, a measurement on the state (5.2) yields the spin $+\frac{1}{2}$ with probability $\frac{2}{3}|Y_{1,0}(\theta, \phi)|^2$ and the spin $-\frac{1}{2}$ with probability $\frac{1}{3}|Y_{1,1}(\theta, \phi)|^2$. If we are not concerned with the angular distribution of the photoelectrons (which is described by the spherical harmonics) but instead integrate over the solid angle, we obtain, since the Y_{lm} are normalized, $\frac{2}{3} \cdot \frac{1}{2} + \frac{1}{3} \cdot (-\frac{1}{2}) = \frac{1}{6}$ for the expectation value of the spin component in the z direction. Correspondingly, the expectation value of the z component of the orbital angular momentum is $\frac{2}{3} \cdot 0 + \frac{1}{3} \cdot 1 = \frac{1}{3}$.

These values can also be obtained from the vector model, as shown in Fig. 5.3b. (One must not, of course, directly form the projections of l and s in the z direction from the arbitrary positions shown. One must first form the projections in the fixed j direction. The time-averaged values of l and s thus obtained are then projected on the z axis).

The situation in the state $^2P_{1/2}$ is quite analogous. For $m_j = \frac{1}{2}$ one has the eigenfunction[3]

$$\sqrt{\frac{1}{3}}\begin{pmatrix}1\\0\end{pmatrix}Y_{1,0} - \sqrt{\frac{2}{3}}\begin{pmatrix}0\\1\end{pmatrix}Y_{1,1}. \tag{5.3}$$

Its representation in the vector model is given in Fig. 5.3c.

What can be learned about the polarization of the photoelectrons from the eigenfunctions just established? First of all we see that for transition 1 (Fig. 5.2) no change occurs in the spin direction since m_s is $+\frac{1}{2}$ in both the initial and final states. For transitions 2 and 3, however, which start from $m_s = -\frac{1}{2}$, there is a definite possibility of a spin flip to $m_s = +\frac{1}{2}$. To determine the fractions of the photoelectrons with $m_s = +\frac{1}{2}$ and $-\frac{1}{2}$ and thus the electron polarization, we must know the probabilities with which the various transitions occur. They are determined by the dipole matrix elements.

To calculate the matrix elements one needs the complete wave functions including their radial parts which are not known exactly. We denote the radial parts by $F(r)$, $F_1(r)$, and $F_3(r)$ which refer to the ground state, $^2P_{1/2}$ state, and $^2P_{3/2}$ state, respectively. $F_1(r)$ is generally different from $F_3(r)$ since the radial parts of the Hamiltonians that result in the $P_{1/2}$ or $P_{3/2}$ states differ in the sign of the spin-orbit coupling potential $(\frac{1}{2}m^{-2}c^{-2})(1/r)(dV/dr)(\mathbf{l}\cdot\mathbf{s})$, the scalar product $(\mathbf{l}\cdot\mathbf{s})$ being negative for $j = \frac{1}{2}$ and positive for $j = \frac{3}{2}$ (see Figs. 5.3b and 5.3c, and Sect. 3.2). Since the radial parts of the Hamiltonians differ, one obtains different radial eigenfunctions. The difference between $F_1(r)$ and $F_3(r)$ increases with increasing spin-orbit coupling, i.e., with increasing atomic number.

Using the abbreviations $\alpha = \begin{pmatrix}1\\0\end{pmatrix}$, $\beta = \begin{pmatrix}0\\1\end{pmatrix}$, $R_{1,3} = \langle F_{1,3}(r)|r|F(r)\rangle$, and the dipole operator $x + iy$ of circularly polarized σ^+ light, the matrix elements for transitions 1, 2, 3 in Fig. 5.2 are

$$b_1 = \langle \varepsilon\,^2P_{3/2}, m_j = \tfrac{3}{2}|x + iy|\,^2S_{1/2}, m_j = \tfrac{1}{2}\rangle$$
$$= \langle F_3(r)\alpha Y_{1,1}|x + iy|F(r)\alpha Y_{0,0}\rangle = -\sqrt{\tfrac{2}{3}}\,R_3 \tag{5.4}$$

---

[3] cf. footnote 4 on page 103.

$$b_2 = \langle \varepsilon\,^2P_{1/2}, m_j = \tfrac{1}{2} | x + iy |\,^2S_{1/2}, m_j = -\tfrac{1}{2}\rangle$$
$$= \langle F_1(r)(\sqrt{\tfrac{1}{3}}\alpha Y_{1,0} - \sqrt{\tfrac{2}{3}}\beta Y_{1,1}) | x + iy | F(r)\beta Y_{0,0}\rangle$$
$$= +\tfrac{2}{3} R_1 \tag{5.5}$$

$$b_3 = \langle \varepsilon\,^2P_{3/2}, m_j = \tfrac{1}{2} | x + iy |\,^2S_{1/2}, m_j = -\tfrac{1}{2}\rangle$$
$$= \langle F_3(r)(\sqrt{\tfrac{2}{3}}\alpha Y_{1,0} + \sqrt{\tfrac{1}{3}}\beta Y_{1,1}) | x + iy | F(r)\beta Y_{0,0}\rangle$$
$$= -\frac{\sqrt{2}}{3} R_3. \tag{5.6}$$

The integrations carried out above are very simple as only the lowest spherical harmonics (which are merely sine and cosine functions) occur. In addition, the orthogonality of the spin functions α and β has been used (see Problem 5.2).

The wave function of the electrons that have made the transitions 2 and 3 starting from the same level is given by the coherent superposition

$$b_2 \cdot |\varepsilon\,^2P_{1/2}, m_j = \tfrac{1}{2}\rangle + b_3 \cdot |\varepsilon\,^2P_{3/2}, m_j = \tfrac{1}{2}\rangle \tag{5.7}$$

(common factors which cancel out in the calculation of the polarization, such as the intensity of the ionizing light, have been omitted). For transition 1, which describes the ionization of a different atom of our incoherent mixture of A↑ and A↓, the corresponding expression is

$$b_1 \cdot |\varepsilon\,^2P_{3/2}, m_j = \tfrac{3}{2}\rangle. \tag{5.8}$$

By substituting (5.1) to (5.6) into (5.7) and (5.8) and rearranging according to spin functions, we obtain

$$-\frac{1}{3}\sqrt{\frac{2}{3}}[\sqrt{2}(R_3 - R_1)Y_{1,0}\alpha + (2R_1 + R_3)Y_{1,1}\beta]$$
$$= -\frac{1}{3}\sqrt{\frac{2}{3}}\begin{pmatrix}\sqrt{2}(R_3 - R_1)Y_{1,0} \\ (2R_1 + R_3)Y_{1,1}\end{pmatrix} \tag{5.9}$$

in the case of transitions 2 and 3 for the part of the wave function which determines the angular momenta. In the case of transition 1 we obtain

$$-\sqrt{\frac{2}{3}}\begin{pmatrix}R_3 Y_{1,1} \\ 0\end{pmatrix}. \tag{5.10}$$

According to Section 2.3, the density matrices of the final states are, from (5.9) and (5.10),

$$\rho_{2+3} = C \begin{pmatrix} 2(R_3 - R_1)^2|Y_{1,0}|^2 & \sqrt{2}(R_3 - R_1)(2R_1 + R_3)Y_{1,0}Y_{1,1}^* \\ \sqrt{2}(R_3 - R_1)(2R_1 + R_3)Y_{1,0}^*Y_{1,1} & (2R_1 + R_3)^2|Y_{1,1}|^2 \end{pmatrix} \quad (5.11)$$

$$\rho_1 = C \begin{pmatrix} 9R_3^2|Y_{1,1}|^2 & 0 \\ 0 & 0 \end{pmatrix}, \quad (5.12)$$

where common factors that are not important in our considerations have been expressed by the constant C.

These density matrices can be used to calculate the electron polarization that arises in the photoionization of the A↓ and A↑ beams. To calculate the polarization of the electron mixture which arises from the photoionization of the unpolarized atomic beam, we form the density matrix of the mixed state, which according to (2.22) is the sum of the matrices (5.11) and (5.12):

$$\rho = C \begin{pmatrix} 9R_3^2|Y_{1,1}|^2 + 2(R_3-R_1)^2|Y_{1,0}|^2 & \sqrt{2}(R_3-R_1)(2R_1+R_3)Y_{1,0}Y_{1,1}^* \\ \sqrt{2}(R_3 - R_1)(2R_1 + R_3)Y_{1,0}^*Y_{1,1} & (2R_1 + R_3)^2|Y_{1,1}|^2. \end{pmatrix} \quad (5.13)$$

Since from (2.21) one has $P_i = \text{tr}\,\rho\sigma_i / \text{tr}\,\rho$ for the components of the polarization vector, one obtains, if one chooses the z component as an example, the angle-dependent expression

$$P_z = \frac{9R_3^2|Y_{1,1}|^2 + 2(R_3 - R_1)^2|Y_{1,0}|^2 - (2R_1 + R_3)^2|Y_{1,1}|^2}{9R_3^2|Y_{1,1}|^2 + 2(R_3 - R_1)^2|Y_{1,0}|^2 + (2R_1 + R_3)^2|Y_{1,1}|^2}. \quad (5.14)$$

Using the rearrangement made in Problem 5.3, one obtains

$$P_z = \frac{\tfrac{9}{2}R_3^2 \sin^2\theta + 2(R_3 - R_1)^2 \cos^2\theta - \tfrac{1}{2}(2R_1 + R_3)^2 \sin^2\theta}{2(R_3 - R_1)^2 + (6R_1R_3 + 3R_3^2)\sin^2\theta}$$

$$= \frac{2(R_3 - R_1)^2 + 2(R_3^2 + R_1R_3 - 2R_1^2)\sin^2\theta}{2(R_3 - R_1)^2 + (6R_1R_3 + 3R_3^2)\sin^2\theta}. \quad (5.15)$$

In the polarization experiments performed so far, all photoelectrons were collected regardless of their direction of emission. We are therefore interested in the polarization $\bar{P}$ averaged over all angles. In this averaging, the polarization values must be weighted with the corresponding intensities I. According to Problem 5.3, $\bar{P}_x$ and $\bar{P}_y$ vanish. Thus

$$\bar{P} = \bar{P}_z = \frac{\iint I P_z \sin\theta d\theta d\phi}{\iint I \sin\theta d\theta d\phi}. \tag{5.16}$$

From the form $P_z = (N_\uparrow - N_\downarrow)/(N_\uparrow + N_\downarrow) = \operatorname{tr} \rho\sigma_z/\operatorname{tr} \rho$ of the polarization formula it can be seen that $\operatorname{tr} \rho$ is proportional to the intensity of the photoelectrons. Therefore, with (5.14) and (5.16), and using the fact that the spherical harmonics are normalized, one obtains

$$\bar{P}_z = \frac{9R_3^2 + 2(R_3 - R_1)^2 - (2R_1 + R_3)^2}{9R_3^2 + 2(R_3 - R_1)^2 + (2R_1 + R_3)^2} = \frac{1 + 2X}{2 + X^2} \tag{5.17}$$

with

$$X = \frac{2R_3 + R_1}{R_3 - R_1}. \tag{5.18}$$

Problem 5.1: Show that $\begin{pmatrix}1\\0\end{pmatrix} Y_{1,1}$ is an eigenfunction of j^2 and j_z with the respective eigenvalues $j(j+1)\hbar^2 = 3/2 \cdot (5/2)\hbar^2$ and $m_j\hbar = (3/2)\hbar$. Show the analogous relations for the state $|{}^2P_{3/2}, m_j = \frac{1}{2}\rangle$.

Solution: Using the abbreviations $\alpha = \begin{pmatrix}1\\0\end{pmatrix}$, $\beta = \begin{pmatrix}0\\1\end{pmatrix}$ we have

$$j^2 \alpha Y_{1,1} = (l^2 + s^2 + 2l \cdot s)\alpha Y_{1,1} = [(2 + \tfrac{3}{4})\hbar^2 + 2l \cdot s]\alpha Y_{1,1}.$$

With the relations following from (2.2)

$$\sigma_x \alpha = \beta,\ \sigma_y \alpha = i\beta,\ \sigma_z \alpha = \alpha,\ \sigma_x \beta = \alpha,\ \sigma_y \beta = -i\alpha,\ \sigma_z \beta = -\beta$$

one has

$$2l \cdot s \alpha Y_{1,1} = \hbar(l_x \sigma_x + l_y \sigma_y + l_z \sigma_z)\alpha Y_{1,1} = \hbar(l_x \beta + il_y \beta + \hbar\alpha)Y_{1,1}.$$

From the following relations which can be found in textbooks on quantum mechanics

$$(l_x + il_y)Y_{lm} = \hbar\sqrt{(l-m)(l+m+1)}\,Y_{lm+1}\quad[4]$$
$$(l_x - il_y)Y_{lm} = \hbar\sqrt{(l+m)(l-m+1)}\,Y_{lm-1}, \tag{5.19}$$

one has $(l_x + il_y)Y_{1,1} = 0$. Thus it follows that

$$j^2 \alpha Y_{1,1} = \left(2 + \frac{3}{4} + 1\right)\hbar^2 \alpha Y_{1,1} = \frac{3}{2} \cdot \frac{5}{2} \hbar^2 \alpha Y_{1,1}.$$

---

[4] Variations in the sign in various publications arise from differing definitions of Y_{lm}. We use the notation in which $Y_{l,-m} = (-1)^m Y_{lm}$.

5. Polarized Electrons by Ionization

In addition one has

$$j_z \alpha Y_{1,1} = (l_z + s_z) \alpha Y_{1,1} = \left(\hbar + \frac{\hbar}{2}\right) \alpha Y_{1,1} = \frac{3}{2} \hbar \alpha Y_{1,1}.$$

For $|{}^2P_{3/2}, m_j = \frac{1}{2}\rangle$ it follows that

$$j^2 \sqrt{\frac{1}{3}} (\sqrt{2} \alpha Y_{1,0} + \beta Y_{1,1})$$

$$= [l^2 + s^2 + \hbar(l_x \sigma_x + l_y \sigma_y + l_z \sigma_z)] \sqrt{\frac{1}{3}} (\sqrt{2} \alpha Y_{1,0} + \beta Y_{1,1})$$

$$= \hbar^2 \left(2 + \frac{3}{4}\right) \sqrt{\frac{1}{3}} (\sqrt{2} \alpha Y_{1,0} + \beta Y_{1,1}) + \hbar(l_x \beta + il_y \beta) \sqrt{\frac{2}{3}} Y_{1,0}$$

$$+ \hbar(l_x \alpha - il_y \alpha - \hbar \beta) \sqrt{\frac{1}{3}} Y_{1,1}.$$

From the last two terms, together with (5.19), one has

$$\frac{2}{\sqrt{3}} \hbar^2 \beta Y_{1,1} + \sqrt{\frac{2}{3}} \hbar^2 \alpha Y_{1,0} - \sqrt{\frac{1}{3}} \hbar^2 \beta Y_{1,1}.$$

Thus one obtains

$$j^2 \sqrt{\frac{1}{3}} (\sqrt{2} \alpha Y_{1,0} + \beta Y_{1,1})$$

$$= \hbar^2 \left(2 + \frac{3}{4}\right) \sqrt{\frac{1}{3}} (\sqrt{2} \alpha Y_{1,0} + \beta Y_{1,1}) + \sqrt{\frac{1}{3}} \hbar^2 \beta Y_{1,1} + \sqrt{\frac{2}{3}} \hbar^2 \alpha Y_{1,0}$$

$$= \frac{3}{2} \cdot \frac{5}{2} \hbar^2 \sqrt{\frac{1}{3}} (\sqrt{2} \alpha Y_{1,0} + \beta Y_{1,1}).$$

In addition one has

$$j_z \sqrt{\frac{1}{3}} (\sqrt{2} \alpha Y_{1,0} + \beta Y_{1,1}) = \sqrt{\frac{1}{3}} \left[\sqrt{2}\left(0 + \frac{\hbar}{2}\right) \alpha Y_{1,0} + \left(\hbar - \frac{\hbar}{2}\right) \beta Y_{1,1}\right]$$

$$= \frac{\hbar}{2} \sqrt{\frac{1}{3}} (\sqrt{2} \alpha Y_{1,0} + \beta Y_{1,1}).$$

Problem 5.2: Calculate the value of b_2 given in (5.5).

Solution: Since $x = r \sin \theta \cos \phi$, $y = r \sin \theta \sin \phi$, and $\langle \alpha | \beta \rangle = 0$, $\langle \beta | \beta \rangle = 1$, it follows that

$$b_2 = -\left\langle F_1(r) \sqrt{\frac{2}{3}} Y_{1,1} \middle| r \sin \theta e^{i\phi} \middle| F(r) Y_{0,0} \right\rangle$$

$$= \sqrt{\frac{2}{3}} R_1 \left(\sqrt{\frac{3}{8\pi}}\right) \int_0^\pi \int_0^{2\pi} \sin \theta e^{-i\phi} \sin \theta e^{i\phi} \sqrt{\frac{1}{4\pi}} \sin \theta d\theta d\phi$$

$$= \frac{2\pi}{4\pi} R_1 \int_0^\pi \sin^3 \theta d\theta = \frac{2}{3} R_1.$$

Problem 5.3: Calculate P_x and show that the average values $\bar{P}_x$ and $\bar{P}_y$ vanish

$$\left(\text{use } Y_{1,0} = Y_{1,0}^* = \sqrt{\frac{3}{4\pi}} \cos \theta, \; Y_{1,1} = -\sqrt{\frac{3}{8\pi}} \sin \theta e^{i\phi}\right).$$

Solution: With ρ from (5.13) we obtain

$$\begin{aligned}
P_x &= \frac{\mathrm{tr}\,\rho \begin{pmatrix} 0 & 1 \\ 1 & 0 \end{pmatrix}}{\mathrm{tr}\,\rho} \\
&= \frac{\sqrt{2}(R_3 - R_1)(2R_1 + R_3)(Y_{1,0} Y_{1,1}^* + Y_{1,0}^* Y_{1,1})}{9R_3^2|Y_{1,1}|^2 + 2(R_3 - R_1)^2|Y_{1,0}|^2 + (2R_1 + R_3)^2|Y_{1,1}|^2} \\
&= \frac{\sqrt{2}(R_3 - R_1)(2R_1 + R_3) Y_{1,0}(Y_{1,1}^* + Y_{1,1})}{2(R_3 - R_1)^2|Y_{1,0}|^2 + (4R_1^2 + 4R_1 R_3 + 10R_3^2)|Y_{1,1}|^2} \\
&= -\frac{(R_3 - R_1)(2R_1 + R_3)\cos\theta\sin\theta\,2\cos\phi}{2(R_3 - R_1)^2(1 - \sin^2\theta) + (2R_1^2 + 2R_1 R_3 + 5R_3^2)\sin^2\theta} \\
&= -\frac{2(R_3 - R_1)(2R_1 + R_3)\cos\theta\sin\theta\cos\phi}{2(R_3 - R_1)^2 + (6R_1 R_3 + 3R_3^2)\sin^2\theta}.
\end{aligned} \quad (5.20)$$

It can immediately be seen from the first term that

$$\int\!\int I P_x \sin\theta\,d\theta\,d\phi$$
$$= \sqrt{2}(R_3 - R_1)(2R_1 + R_3) \int\!\int (Y_{1,0} Y_{1,1}^* + Y_{1,0}^* Y_{1,1}) \sin\theta\,d\theta\,d\phi$$

vanishes since the spherical harmonics are orthogonal. For $\bar{P}_y$ it is unnecessary to make the analogous calculation: Since in the arrangement discussed we have rotational symmetry, the x and y directions are equivalent. Because $\bar{P}_x$ is 0, $\bar{P}_y$ must also be 0.

5.2.2 Illustration of the Fano Effect. Experimental Results

According to the preceding formulae, the attainable polarization is determined by the radial matrix elements R_1 and R_3. It can be seen immediately that for $R_1 = R_3$ (vanishing spin-orbit coupling) the polarization vanishes. It is therefore not obvious that polarized photoelectrons arise from ionization with circularly polarized light (i.e., spin-oriented photons). The occurrence of electron polarization cannot simply be inferred from the balance of angular momentum: With vanishing spin-orbit coupling, the photon spin is transferred to the orbital angular momentum of the photoelectron, as described by the selection rule $\Delta m_l = \pm 1$ for circularly polarized light. Only if there is an interaction between spin and orbital angular momentum of the photoelectron can the preferential orientation of the photon spins lead to a preferential orientation of the electron spins.

To avoid getting lost in the calculations we shall try to illustrate how the Fano effect arises. First we shall consider the electron polarization which arises from photoionization of A↓, i.e., from transitions 2 and 3 in Fig. 5.2. We must then start from ρ_{2+3} [(5.11)] and instead of (5.17) which refers to photoionization of an unpolarized atomic beam we obtain

$$\bar{P}_z = \frac{2(R_3 - R_1)^2 - (2R_1 + R_3)^2}{2(R_3 - R_1)^2 + (2R_1 + R_3)^2} = \frac{N_\uparrow - N_\downarrow}{N_\uparrow + N_\downarrow} = \frac{Q_e^\uparrow - Q_e^\downarrow}{Q_e^\uparrow + Q_e^\downarrow}. \quad (5.21)$$

$Q_e^\uparrow$ and $Q_e^\downarrow$ are the cross sections for photoproduction of e↑ and e↓ for the target being considered. If the spin-orbit coupling vanishes ($R_1 = R_3$) one obtains $\bar{P}_z = -1$. This is to be expected since the atomic beam is totally polarized in the $-z$ direction and the electron spin is not affected by the photoionization if there is no spin-orbit interaction. If $R_3 \neq R_1$ we see from (5.21) that $Q_e^\uparrow \neq 0$. This means that some of the spins will flip during photoionization and reduce the degree of polarization arising in these transitions.

The polarization of the photoelectrons produced by transition 1 can be calculated from (5.12):

$$\bar{P}_z = \frac{Q_e^\uparrow - Q_e^\downarrow}{Q_e^\uparrow + Q_e^\downarrow} = \frac{9R_3^2 - 0}{9R_3^2 + 0} = 1. \tag{5.22}$$

In this case no spin flips occur: $Q_e^\downarrow$ is always zero.

Thus we see that the spin-orbit interaction has two effects:
a) it may cause a spin flip in transitions 2 and 3;
b) it leads to differing cross sections for the photoionization of A↑ and A↓:

$$\frac{Q(\text{A}\downarrow)}{Q(\text{A}\uparrow)} = \frac{2(R_3 - R_1)^2 + (2R_1 + R_3)^2}{9R_3^2} \tag{5.23}$$

[$Q(\text{A}\downarrow)/Q(\text{A}\uparrow)$ is the ratio of the denominators in (5.21) and (5.22), which refer to the photoionization of A↓ and A↑, respectively]. With vanishing spin-orbit coupling this ratio is 1.

The resulting polarization is determined by these two effects of spin-orbit coupling. It is quite easy to see that the average polarization $\bar{P}_z$ is not necessarily parallel to the spin direction of the incident photons. If, for example, $R_1 = 4R_3$ then from (5.23) $Q(\text{A}\downarrow)/Q(\text{A}\uparrow) = 11$ and from (5.21) $Q_e^\uparrow/Q_e^\downarrow = 2(R_3 - R_1)^2/(2R_1 + R_3)^2 = 2/9$. This means that the probability for the A↓ to be ionized is 11 times greater than that for the A↑, and that most of the spin directions are retained. Therefore a negative polarization arises as also follows from (5.17): $\bar{P}_z = -\tfrac{1}{2}$.

If, on the other hand, $R_3 = -2R_1$, then (5.21) yields $Q_e^\downarrow = 0$. This means that in the ionization of the A↓, all spins must flip into the $+z$ direction. We then obtain only photoelectrons with spins parallel to the z direction. Correspondingly, (5.17) yields $\bar{P}_z = 1$. This can also be explained as follows: In forming the expression (5.9) from $|\varepsilon^2 P_{1/2}, m_j = \tfrac{1}{2}\rangle$ and $|\varepsilon^2 P_{3/2}, m_j = \tfrac{1}{2}\rangle$, terms containing the spin-down functions cancel due to interference, if $R_3 = -2R_1$. In this exceptional case, $m_s = +\tfrac{1}{2}$

and $m_l = 0$ are good quantum numbers for the state resulting from transitions 2 and 3.

Since spin-orbit coupling is only a small interaction, such large polarization effects cannot usually be expected. Normally the polarization does not differ significantly from 0, the value without spin-orbit coupling. The circumstances which favor high degrees of polarization can be illustrated by the following picture which in fact led to Fano's considerations: The photoionization cross sections, together with the transition matrix elements —and thus also the polarization of the photoelectrons—depend on the wavelength of the incident light. For most alkali atoms this dependence is particularly pronounced in an experimentally convenient region: The photoionization cross section $Q_e = Q_e^\uparrow + Q_e^\downarrow$ passes through a deep minimum near the threshold. Due to the spin-orbit interaction we obtain slightly different cross sections $Q_e^\uparrow$ and $Q_e^\downarrow$ for the photoproduction of $e\uparrow$ and $e\downarrow$ (see Fig. 5.4) since one obtains from (5.17)

$$\frac{Q_e^\uparrow}{Q_e^\downarrow} = \frac{9R_3^2 + 2(R_3 - R_1)^2}{(2R_1 + R_3)^2}. \tag{5.24}$$

Although the small spin-orbit interaction does not generally lead to large

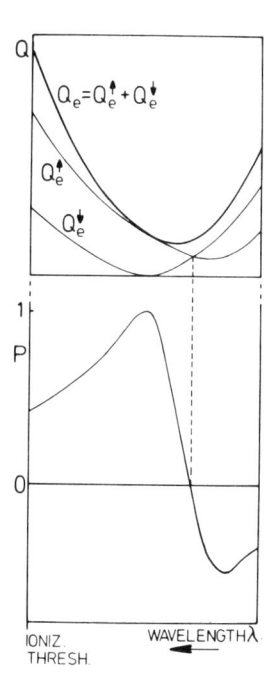

Fig. 5.4. Qualitative diagram of the photoionization cross section and the photoelectron polarization as a function of wavelength for alkali atoms ($Q_e^\downarrow = 0$ for $R_3 = -2R_1$)

differences between the two cross sections, their ratios near the minima are quite considerable.[5] If one uses wavelengths near the minima, one obtains electrons in predominantly one spin state, i.e., a high degree of polarization. The situation is therefore quite analogous to that in electron scattering, which was discussed in Subsection 3.4.2: The shape of the polarization curves is determined by the cross-section curves for producing e↑ and e↓.

To be able to quantitatively determine the curves given in Fig. 5.4, R_1 and R_3 must be known. It is, however, difficult to calculate the radial parts of the wave functions and thus the transition matrix elements with enough precision that even the difference $R_3 - R_1$ occurring in the formulae is still reliable. Fano therefore tried to evaluate the wavelength dependence $\bar{P}_z(\lambda)$ by making use of the fact that the spin-orbit interaction responsible for the polarization effect also has other consequences which had been studied earlier: Apart from the doublet splitting of the alkali energy levels, spin-orbit interaction determines the intensity ratio of the doublet lines and the exact shape of the photoionization cross section Q_e (even if the spin direction of the photoelectrons is not taken into account). From such data he estimated the parameter X—connected with R_1 and R_3 by (5.18)— as a function of the wavelength λ. He predicted approximately the results that were obtained by the measurements we will now discuss.

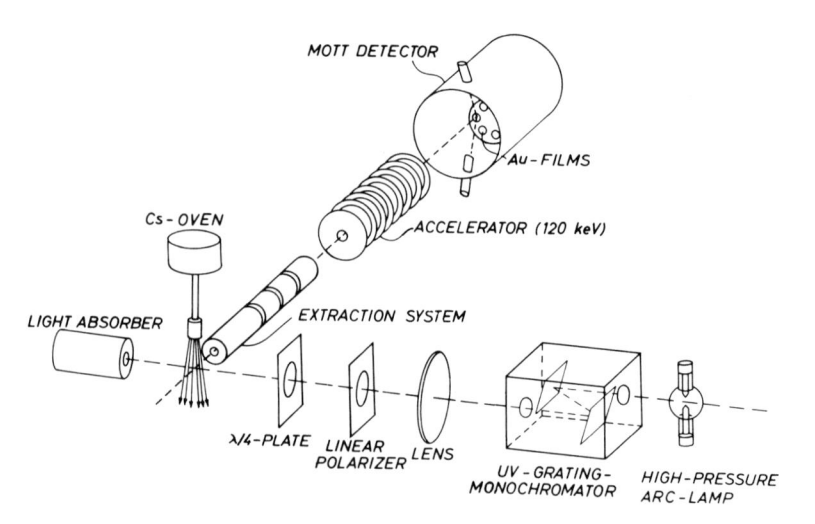

Fig. 5.5. Experimental arrangement for measuring the Fano effect [5.5]

---

[5] The strong deviation of the ratio R_1/R_3 from 1 in the numerical example above can only occur near zeros of R_1 and R_3 because the absolute difference between these quantities is very small due to the small spin-orbit coupling. Since the photoionization cross sections are determined by R_1 and R_3, the minima of the cross sections and the zeros of R_1 and R_3 lie in the same wavelength range.

5.2 Fano Effect

Figure 5.5 is a schematic diagram of an experimental setup for measuring the Fano effect [5.5]. An unpolarized cesium-vapor beam is crossed by monochromatic circularly polarized uv light. The photoelectrons produced are collected by an extraction system, irrespective of their direction of emission. The subsequent measurement of their polarization which is carried out with a Mott detector (see Subsect. 3.6.3) thus yields the average value $\bar{P}_z$, for brevity denoted as P in the following. The task of suppressing the numerous background electrons that come from the chamber walls or other parts of the apparatus is accomplished by suitable electron-optical potential barriers. In addition, it must be ensured, through the choice of a suitable oven temperature, that the portion of the Cs_2 molecules in the Cs beam is small: The high photoionization cross section of these dimers must not give rise to an appreciable number of unwanted photoelectrons.

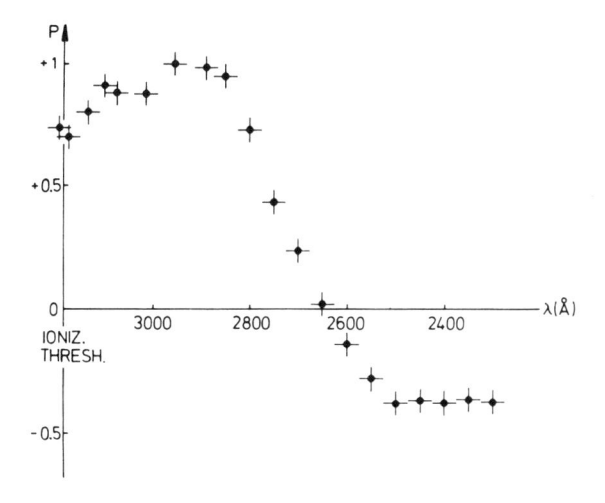

Fig. 5.6. Experimental values of the photoelectron polarization vs. wavelength for the Fano effect with cesium [5.5]

Figure 5.6 shows the results of these measurements which were made in the interesting wavelength region where the cross-section minimum occurs. It can be seen that at 2900 Å total polarization is achieved within the limits of experimental accuracy. This makes the Fano effect of interest as a source of polarized electrons. In addition, such measurements yield information on the parameter $X(\lambda)$ from (5.18) and thus on the radial matrix elements. In this way one obtains more accurate information than was previously available on the influence of spin-orbit coupling on the aforementioned properties of the alkali atoms [5.6–8].

Although the wavelength dependence of the photoelectron polarization has not been measured for alkali atoms other than Cs (for solid alkalis, see Sect. 6.2), it can easily be drawn from an equivalent experiment which has been made with K, Rb and Cs. A↑ and A↓ have been separately photoionized by circularly polarized light [5.6]. The ratio of the photoionization cross sections $Q(A\uparrow)/Q(A\downarrow)$ is given by (5.23). Comparison with (5.24) shows that the information obtained from a measurement of $Q(A\uparrow)/Q(A\downarrow)$ is equivalent to that obtained from measuring the polarization $P = (Q_e^\uparrow - Q_e^\downarrow)/(Q_e^\uparrow + Q_e^\downarrow) = [(Q_e^\uparrow/Q_e^\downarrow) - 1]/[(Q_e^\uparrow/Q_e^\downarrow) + 1]$. We can therefore say that there is good quantitative knowledge of the Fano effect, not only for Cs, but also for Rb and K (see also [7.42, 43]).

The angular dependence of the photoelectron polarization [examples in (5.15) and (5.20)] has been studied in several theoretical papers [5.9–11], but experimental results are not yet available. They would make it possible to completely determine the essential parameters characterizing the photoionization process [5.11]. It is interesting to note that photoelectrons ejected into an arbitrary direction from certain unpolarized targets by unpolarized light should be polarized perpendicular to the reaction plane.[6] This is an analogy to electron scattering, where electrons polarized perpendicular to the scattering plane are produced by scattering of an unpolarized beam by an unpolarized target. The polarization averaged over all directions is, of

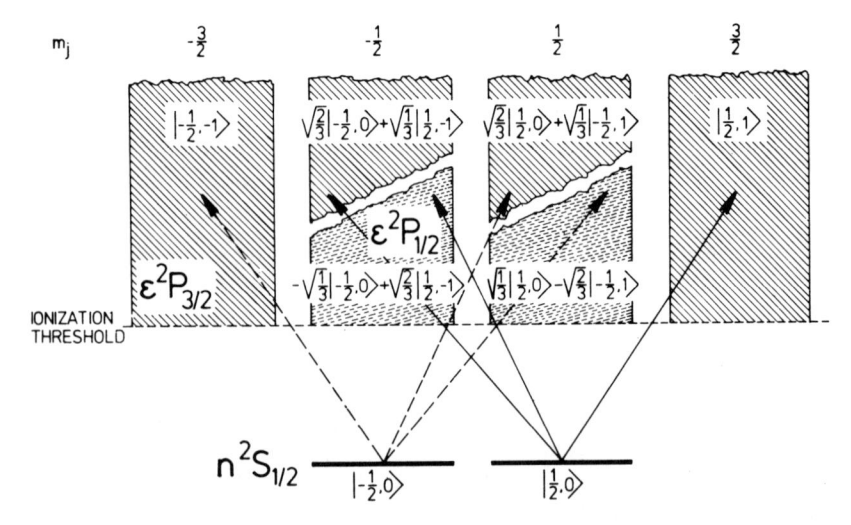

Fig. 5.7. Photoionization of alkali atoms with linearly polarized light (transitions denoted by arrows of the same kind must be superposed coherently, other transitions must be superposed incoherently)

---

[6] Defined by incident photon and outgoing electron.

course, zero in both cases. The last remarks imply that here is another area of polarized-electron physics in which there are quite a few interesting experiments to be done.

Problem 5.4: Show without calculation that with linearly polarized light and unpolarized atoms one obtains $\bar{P}_z = 0$ in the photoionization process discussed in this section.

Solution: The linearly polarized light incident in the z direction can be considered as a coherent superposition of right- and left-circularly polarized light. Apart from the transitions with $\Delta m_j = +1$, those with $\Delta m_j = -1$ also take place. The orientations of corresponding angular momenta on the left and on the right side of Fig. 5.7 are mutually opposite. When the final states reached from A↑ or A↓ are incoherently superposed (shown respectively with solid and dashed arrows in Fig. 5.7) one finds for each polarization state a state with the opposite polarization and the same weight. The resulting polarization P_z is thus zero.

5.3 Autoionizing Transitions

The polarization of photoelectrons from atoms other than alkalis may be considerably influenced by autoionization resonances. The resonance behavior of the spin polarization is discussed using the example of thallium atoms.

The conspicuous polarization effects obtained with alkali atoms prompt the question: Can similar results be obtained with other atoms, too? Recent theoretical work answers this question in the affirmative [5.9–12]. In discussing the photoelectron polarization in these cases we must take into account the fact that in many atoms autoionizing transitions play a part in the photoionization process. They may cause a resonance structure of the polarization curve as we will now show for the specific example of thallium.

Figure 5.8 shows the relevant energy states of thallium. Due to the selection rules $\Delta l = \pm 1$, $\Delta j = 0, \pm 1$, and $\Delta m_j = +1$ (σ^+ light), the states $\varepsilon^2 S_{1/2}$, $\varepsilon^2 D_{3/2}(m_j = \frac{1}{2})$, and $\varepsilon^2 D_{3/2}(m_j = \frac{3}{2})$ are accessible from the $6^2 P_{1/2}$ ground states of the unpolarized thallium atoms.

The spin polarization of the photoelectrons in a specific final state follows immediately from the coupling coefficients of the wave functions $|m_s, m_l\rangle$ given in Fig. 5.8. For example, we find for the state $\varepsilon^2 D_{3/2}(m_j = \frac{1}{2})$ the polarization $P = (\frac{1}{5} - \frac{4}{5})/(\frac{1}{5} + \frac{4}{5}) = -0.6$ (see Problem 5.5). Superposition of the spin polarizations of the various final states weighted with the corresponding transition probabilities yields the photoelectron polarization. Evaluation of the transition probabilities similar to Problem 5.2 shows that the two final $\varepsilon^2 D_{3/2}$ states together contribute a polarization of -0.5, whereas the $\varepsilon^2 S_{1/2}$ state has a polarization of $+1$.

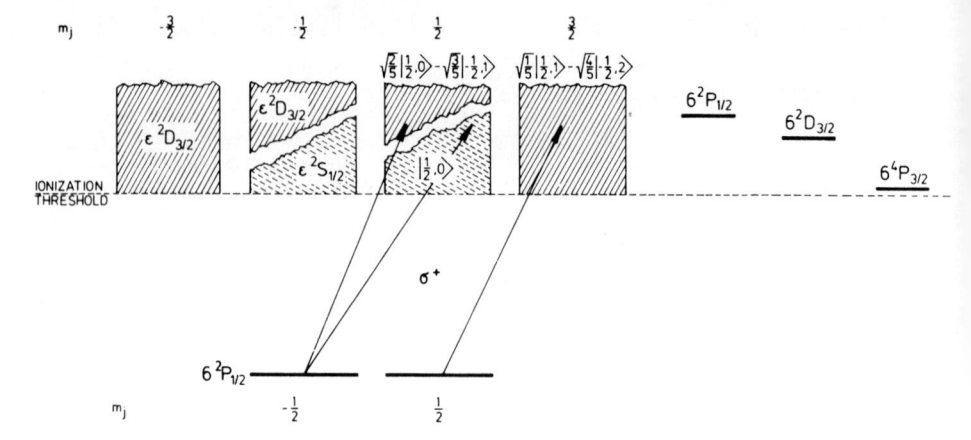

Fig. 5.8. Autoionizing states and continuum states reached from the ground state $6^2P_{1/2}$ of thallium with circularly polarized light

Because of the different signs of these values the resulting polarization

$$P = (1 \cdot Q_S - 0.5 \cdot Q_D)/(Q_S + Q_D) \tag{5.25}$$

would normally not be very large, the exact value depending on the ratio of the cross sections for transitions to the S and D states, Q_S/Q_D, which depends on the wavelength λ.

So far we have not taken into account, however, the autoionizing states of thallium which are also shown in Fig. 5.8. They result from excitation of a 6s electron and decay after a short lifetime, so that eventually the same final states are reached via these indirect transitions. The autoionizing state $6^2P_{1/2}$ decays into the $\varepsilon^2S_{1/2}$ continuum, and $6^2D_{3/2}$ and $6^4P_{3/2}$ into $\varepsilon^2D_{3/2}$, since, due to the conservation of angular momentum, J must be the same before and after the decay. At the excitation wavelengths of the autoionizing states one has a resonance behavior of the photoionization cross section Q as shown in Fig. 5.9 which is based on results of MARR and HEPPINSTALL [5.13] and of BERKOWITZ and CHUPKA [5.14].

Whenever one of the cross sections Q_S and Q_D dominates due to the autoionization resonances, the polarization tends to $+100\%$ or -50%, respectively, according to (5.25). As an example, let us consider the wavelengths where Q_S dominates. The analysis of the cross-section resonances shows that Q_D disappears at approximately 49730 and 57239 cm^{-1}, leaving Q_S alone to contribute to P; in these cases the polarization should be exactly 100%. On the other hand, Q_S dominates near 67137 cm^{-1}. A tendency of the polarization towards 100% is then to be expected. Simi-

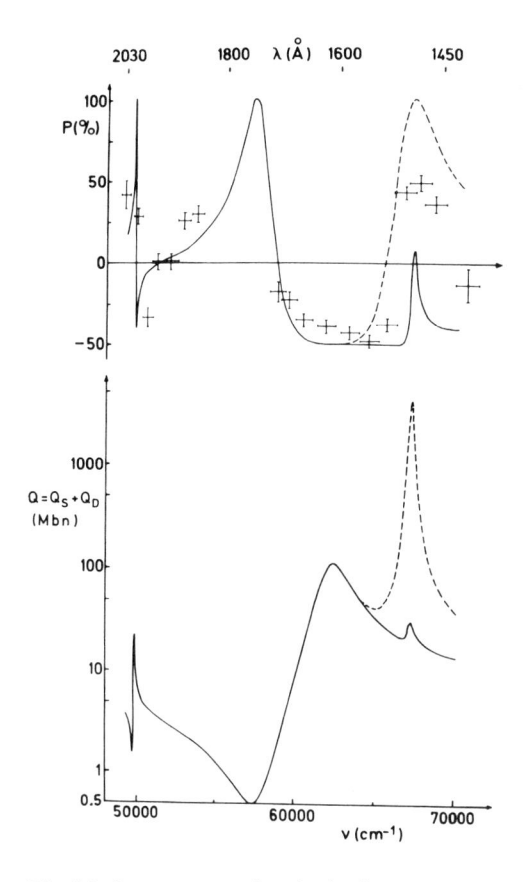

Fig. 5.9. Lower part: photoionization cross section of thallium according to [5.13] (full curve) and [5.14] (broken curve). Upper part: spin polarization of photoelectrons; experimental results and values calculated using the cross sections from the lower part

larly, the negative peaks of P occur at those wavelengths, where Q_D dominates.

Since, according to (5.25), cross-section resonances produce polarization resonances one frequently has polarization peaks located at those wavelengths where the photoionization cross section has a maximum. In the other polarization phenomena discussed so far we often found that the polarization peaks were located near cross-section minima, a fact which renders experimental studies difficult.

Our considerations show that the polarization of the photoelectrons can be predicted on the basis of (5.25) if the relative contributions of Q_S and Q_D to the photoionization cross section can be evaluated. The result of a semi-empirical calculation is seen in the upper part of Fig. 5.9, where the solid and broken lines are the polarization curves based on two different experimental cross-section curves. The figure also shows that experimental studies of this polarization phenomenon have to use wavelengths where it is cumbersome to produce circularly polarized light of sufficient intensity with conventional methods. For polarization studies of photoelectrons from atoms other than alkalis, synchrotron radiation would be an ideal source of polarized uv light.

In the experiment [5.15] sketched in Fig. 5.10 the uv light produced by a H_2 discharge lamp was circularly polarized by a MgF_2 prism in conjunction with a MgF_2 quarter-wave plate and focused onto a thallium-vapor beam emerging from an oven at 1100°C. The photoelectrons were produced in the middle of an asymmetric quadrupole field (not shown) which ensured that all electrons were extracted regardless of their direction of emission and that electrons produced at the walls were not detected. The electron polarization was measured by a highly efficient Mott detector whose I/I_0 ratio was better than 10^{-3} (see Subsect. 3.6.3). It is worth pointing out that for the measurement of $P(\lambda)$ the intensity distribution $I(\lambda)$ of the light need not be known. This is due to the fact that measuring the electron polarization means measuring the *ratio* of the intensities scattered into the two detectors in the Mott chamber, so that knowledge of incident intensities is unnecessary.

The deviation of the measured from the predicted polarization near 1500 Å shows that the photoionization cross section needs to be remeasured here. The measured value $P = -50\%$ at 64400 ± 600 cm^{-1} means that

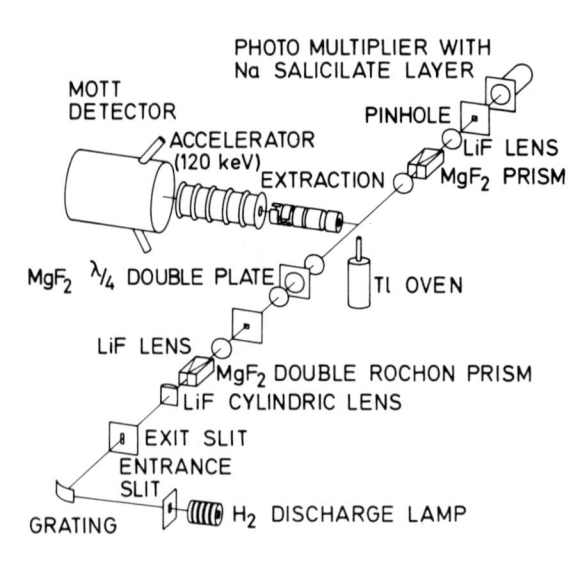

Fig. 5.10. Schematic diagram of the apparatus for measuring the polarization of photoelectrons from thallium [5.15]

at this wave number, Q_S disappears. This special case illustrates the general fact that at each wavelength the polarization measurement determines the relative contributions of the individual cross sections Q_S and Q_D; measurement of the photoionization cross section always yields the sum $Q = Q_S + Q_D$. This is another example of the application of spin-polarization studies. They provide additional information on the autoionization process that is not obtainable from cross-section measurements alone; it can help to classify and understand the autoionization resonances in more complicated cases than discussed here (see also [5.16]).

Problem 5.5: Calculate the spin polarization of the photoelectrons in the state $|\varepsilon^2 D_{3/2}, m_J = 3/2\rangle$ using the eigenfunctions given in Fig. 5.8.

Solution: We are interested in all the photoelectrons regardless of their direction of emission. Since the z axis (propagation direction of the incident light) is the only preferential direction, P_x and P_y vanish when averaged over all angles. Hence we find from the definition of P_z as the expectation value of the spin operator σ_z

$$P \equiv \bar{P} = \bar{P}_z = \langle \sqrt{\tfrac{1}{5}}\binom{1}{0} Y_{2,1} - \sqrt{\tfrac{4}{5}}\binom{0}{1} Y_{2,2} | \sigma_z | \sqrt{\tfrac{1}{5}}\binom{1}{0} Y_{2,1} - \sqrt{\tfrac{4}{5}}\binom{0}{1} Y_{2,2} \rangle$$

$$= \langle \sqrt{\tfrac{1}{5}}\binom{1}{0} Y_{2,1} - \sqrt{\tfrac{4}{5}}\binom{0}{1} Y_{2,2} | \sqrt{\tfrac{1}{5}}\binom{1}{0} Y_{2,1} + \sqrt{\tfrac{4}{5}}\binom{0}{1} Y_{2,2} \rangle.$$

Since the eigenfunctions are orthonormal, one has

$$\bar{P}_z = \tfrac{1}{5} - \tfrac{4}{5} = -0.6.$$

5.4 Multiphoton Ionization

An atom can also by photoionized by absorption of several photons whose total energy is greater than the ionization energy. By using circularly polarized light, one obtains polarized photoelectrons. Cases are discussed where the photoionization takes place via an intermediate state of the atom, i.e., by absorption of a resonance frequency. Numerous possibilities arise for producing polarized electrons.

Until now we have been discussing photoionization that takes place by absorption of one photon. It can, however, also occur by absorption of several photons of lower energy, as long as the sum of the photon energies is greater than the ionization energy. In this case one can work with longer wavelengths and have the advantage of being outside the uv region which, below ca. 1800 Å, presents particular experimental difficulties. If this multiphoton ionization is carried out with circularly polarized light, strongly polarized photoelectrons can be produced, as has been suggested by several authors [5.17–20].

5. Polarized Electrons by Ionization

Multiphoton ionization may occur by excitation of an intermediate resonance level which is subsequently ionized by absorption of another photon. However, when the photon frequencies do not match the energies of any intermediate states the process will also take place, though with lower cross section [5.17]. We will discuss here the first case which is the easiest to realize experimentally.

To be specific, let us consider two-photon ionization of cesium. A photon of resonance frequency excites an intermediate state of the atom which is then ionized by absorption of a second photon. If we illuminate with the wavelength 4593 Å, the intermediate state $7^2P_{1/2}$ is reached (see Fig. 5.11). If σ^+ light is used, the photoionization can only follow route 1 because of the selection rules $\Delta l = \pm 1$, $\Delta j = 0, \pm 1$, and $\Delta m_j = +1$. There are no transitions starting from the ground state with $m_j = +\frac{1}{2}$, nor are there transitions from the excited state $7^2P_{1/2}(m_j = \frac{1}{2})$ into the S states (since $m_j = \frac{3}{2}$ is not possible there) or into the P states of the continuum.

As in Subsection 5.2.1, by using Clebsch-Gordan coefficients or by direct calculation, it can be seen that the final state $\varepsilon^2 D_{3/2}(m_j = 3/2)$ has the angular momentum eigenfunction $\sqrt{\frac{1}{5}}\binom{1}{0}Y_{2,1} - \sqrt{\frac{4}{5}}\binom{0}{1}Y_{2,2}$. Since we are still interested in all photoelectrons regardless of their direction

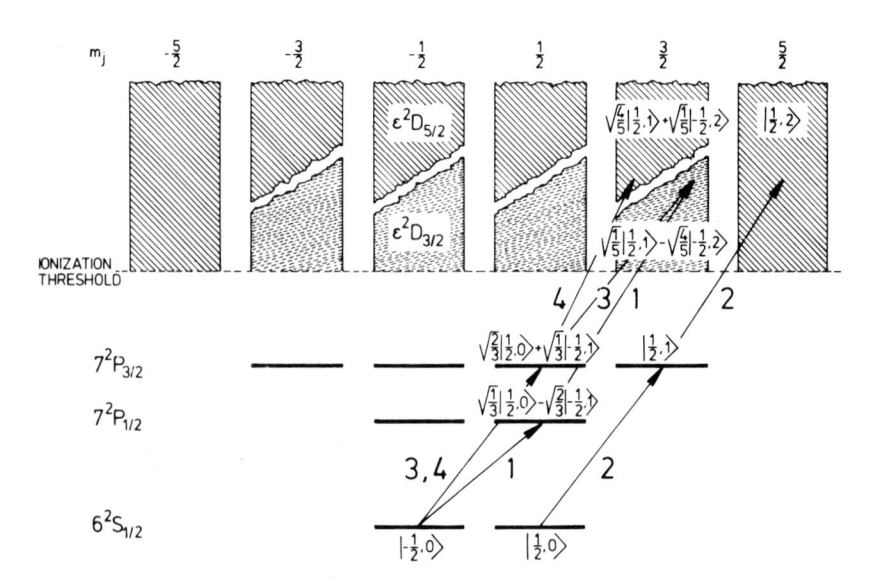

Fig. 5.11. Two-photon transitions in Cs atom

of emission, we calculate their average polarization $\bar{P}_z$ in the direction of the light propagation and obtain $\bar{P}_z = \frac{1}{5} - \frac{4}{5} = -0.6$ (see Problem 5.5). Thus with σ^+ light of wavelength 4593 Å one obtains, by absorption of two photons, a photoelectron polarization of -60%. We have here a different situation than in the case of the Fano effect. There it is essential that, owing to an appreciable spin-orbit coupling in the continuum, the radial matrix elements R_1 and R_3 depend on j. In the present discussion, however, use is made of the energy splitting of the bound states which is likewise caused by spin-orbit coupling: A polarized final state is reached via a polarized intermediate state and dependence of the radial matrix elements on j is not necessary for producing polarization of the photoelectrons.

If one uses σ^+ light of the shorter wavelength 4555 Å one first reaches, by routes 2, 3, and 4, the sublevels $m_j = \frac{1}{2}$ and $\frac{3}{2}$ of the $7^2P_{3/2}$ state and finally the continuum states shown. Superposition of the polarizations of the final states, weighted with the corresponding transition probabilities, yields the resulting photoelectron polarization. Accordingly, the polarization obtained depends in this case on the transition probabilities into the various continuum states. If one assumes that the radial matrix elements to the continuum are equal for $j = \frac{3}{2}$ and $\frac{5}{2}$ (thus excluding the circumstances which cause the Fano effect), a simple calculation similar to that in Subsection 5.2.1 yields $\bar{P}_z = \frac{9}{11} \approx 82\%$. If the radial matrix elements differ, the polarization is somewhat changed.

Only if spin-orbit coupling were to disappear both in the continuum and in the discrete region (radial matrix elements to the continuum not j-dependent, vanishing energy splitting in the discrete region) would the

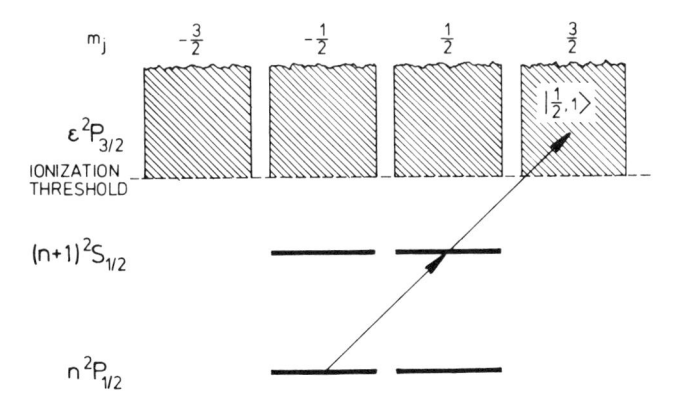

Fig. 5.12. Two-photon transition in trivalent atom

polarization vanish, since in that case the photon spins would not be coupled into the electron spin system.

Multiphoton ionization gives rise to many possibilities for producing polarized electrons if one varies the number of photons participating and the kinds of atoms used. One can [5.18], for example, work with atoms which have the ground state $n^2P_{1/2}$ (B, Al, Ga, In, Tl) and illuminate with the wavelength that excites the intermediate state $(n+1)^2S_{1/2}$ (see Fig. 5.12). Due to the aforementioned selection rules for σ^+ light, the two-photon transition then follows a specific path into the continuum state $\varepsilon^2P_{3/2}(m_j = \frac{3}{2})$ in which all electrons have the spin quantum number $m_s = +\frac{1}{2}$. The photoelectrons are thus totally polarized.

We have dealt with only two of the simplest cases for producing polarized electrons by multiphoton ionization. There are abundant possibilities. By playing around with selection rules, energy levels, and the number of photons, the reader can easily find innumerable other transitions which yield polarized electrons.

Multiphoton ionization can only be achieved with light sources of high intensities. The ionization of short-lived intermediate states, discussed above, only yields measurable intensities if there are enough atoms in these states. With laser beams, it is easily possible to saturate an intermediate state so that there are as many atoms in the intermediate as in the initial state. The required resonance frequencies can be obtained by using tunable dye lasers. The laser can simultaneously produce the photons for the second transition.

Multiphoton ionization is another example of the fact that it does not suffice to discuss the polarization phenomena "in principle". As the light intensity is increased beyond a certain limit, the width of the intermediate resonance level is changed as well as its energy [5.21]. The polarization of the photoelectrons is then affected due to the influence of neighboring levels with different polarizations; this has been theoretically discussed and experimentally verified by preliminary results obtained with sodium [5.22]. Another drastic change of the electron polarization is brought about by the fact that the circular polarization of the light used does not reach the ideal value of 100%. The effect of a small admixture of σ^- light in high-intensity σ^+ light is quite obvious: If one takes no special precautions, there is enough σ^- light in the intense radiation field to saturate the transitions $\Delta m_j = -1$. The subsequent photoionization of the states reached by these unwanted transitions strongly affects the polarization obtained.

We conclude this section by noting that experimental information in this field is particularly sparse. Polarized electrons have been obtained by multiphoton ionization of sodium [5.22] and cesium [5.19]. For a quantitative test of the theory these measurements must, however, be improved.

5.5 Collisional Ionization of Polarized Atoms

By collisional ionization of polarized atoms, polarized free electrons can be produced. The processes discussed in detail are the collision of polarized deuterium atoms with H_2 or He and Penning ionization of polarized helium atoms in a helium discharge. Measurement of the electron polarization can be used as a diagnostic tool for the analysis of collision processes.

5.5.1 Collisional Ionization of Polarized Metastable Deuterium Atoms

If one has polarized atoms one can eject their oriented electrons by ionization and so produce polarized free electrons. This can be done by photoionization, as discussed in Section 5.1, but it can also be done by collisional ionization. Useful degrees of polarization are obtained only if it is mainly the oriented electrons that are ejected in the collisions; ionization of states with non-oriented electrons must be avoided.

In the method [5.23] now to be discussed, polarized deuterium atoms in the state $2^2S_{1/2}$ are used. This state is metastable because radiative transitions into the ground state $1^2S_{1/2}$ are forbidden by the selection rule $\Delta l = \pm 1$. The state lies only 3.4 eV below the ionization threshold and is much easier to ionize than the ground state whose ionization requires about 13.6 eV. When the metastable deuterium atoms collide with H_2 molecules or He atoms, which themselves have high ionization energies, the metastable deuterium atoms are preferentially ionized.

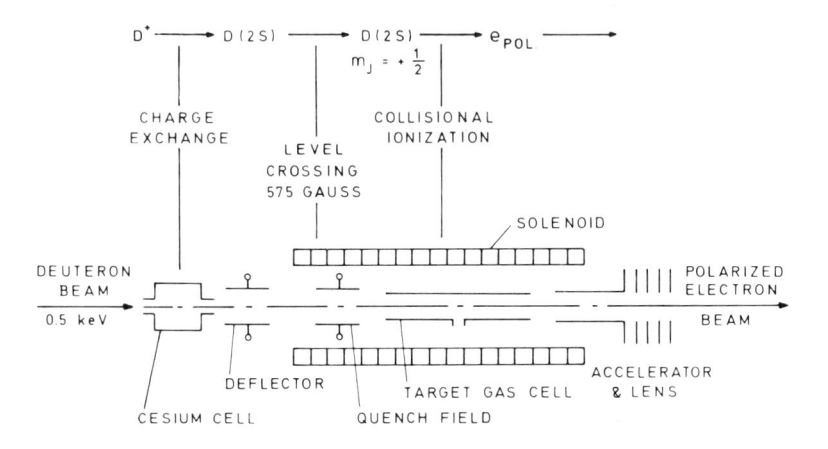

Fig. 5.13. Production and collisional ionization of polarized metastable deuterium atoms [5.23]

The details of the experiment are given in Fig. 5.13. A deuteron beam which passes through Cs vapor picks up electrons by charge exchange. At deuteron energies from 500 to 1000 eV, about 25% of the neutral D atoms emerging from D^+-Cs collisions are in the metastable state $2^2S_{1/2}$. The rest are in states that decay immediately to the deuterium ground state. The remaining charged particles that leave the Cs cell are removed from the beam by a weak electric deflector field.

To polarize the metastable deuterium atoms the "level-crossing" technique is used which takes advantage of the Zeeman splitting in an external magnetic field [5.24]. The $m_j = -\frac{1}{2}$ sublevel of the $2^2S_{1/2}$ state and the $m_j = \frac{1}{2}$ sublevel of the $2^2P_{1/2}$ state cross in a magnetic field of 575 G (see Fig. 5.14). If the states overlap, an external perturbation induces transitions between them without energy input. Accordingly, since atoms in the $2^2P_{1/2}$ state make the optically allowed transition to the ground state within 10^{-9} s, the state $2^2S_{1/2}(m_j = -\frac{1}{2})$ will be completely depopulated. The atoms in the state $2^2S_{1/2}(m_j = +\frac{1}{2})$ remain in this state, so that the arrangement shown in Fig. 5.13 yields longitudinally polarized metastable D atoms. The static magnetic field that gives rise to the Zeeman splitting and that is also strong enough to decouple the electron spins from the nuclear spins may, in principle, be used as the perturbation to bring about the transitions between the overlapping levels. In the present experiment an additional weak electrostatic field ("quench field") is used so that a complete depopulation of the unwanted states really occurs before the atoms reach the ionization cell.

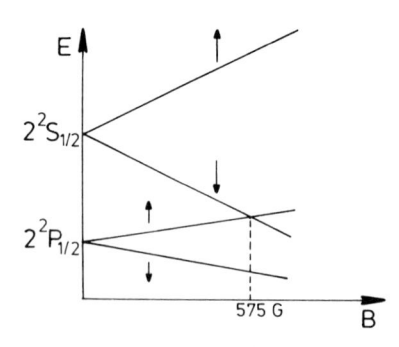

Fig. 5.14. $2^2S_{1/2}$ and $2^2P_{1/2}$ energy levels of atomic deuterium as a function of the magnetic field

The collisional ionization of the metastable D atoms whose spins are oriented parallel to the magnetic field takes place in a gas cell filled with H_2 or He. (To photoionize the atoms would be ineffective here: The photoionization cross section is small, and the D atoms of a few hundred eV have high velocities so that their density in the ionization area is low).

The electrons ejected by collisional ionization are collected by an extraction field. After conversion of their longitudinal polarization into transverse polarization (see Subsect. 3.6.3) the polarization is measured with a Mott detector. The maximum value found was 33%.

Since the ionization cross section of the metastable state is about 15 times larger than that of the ground state and since the ratio of polarized metastable atoms to atoms in the ground state is about 1/8, the polarization ought to be considerably higher than 33% (about 60%) if the polarized bound electrons were directly knocked out during the collision. The fact that the measurement gave $P \leq 33\%$ indicates that the collisional ionization predominantly proceeds as follows: When the polarized D atoms pass through the target gas cell, first D^- ions are formed; they subsequently eject one of their electrons by autoionization. The spins of the electrons that are picked up by the polarized atoms from the unpolarized target gas can be parallel or antiparallel to those of the polarized atoms. Since in the autoionization process there is no difference between the attached electrons and the previously oriented atomic electrons, the observed electron polarization is reduced.

The experimental result thus leads to the conclusion that the collisional ionization of the metastable D atoms does not take place directly, but instead occurs predominantly by autoionization. This is an example of how polarization measurements can yield essential information on atomic collision processes.

5.5.2 Penning Ionization

We will now give another example of the diagnostic possibilities arising from electron-polarization studies: It has been shown by such investigations [5.25] that the free electrons in a weak He discharge are mainly produced by collisions between metastable He atoms (He*) and not by ionization of the atoms through electron collisions, as one might suppose.

Figure 5.15 is a schematic diagram of the experiment. A weak rf discharge produces a steady-state population of the metastable helium states. Because of their long lifetime (fraction of a millisecond) the metastable states are populated to a much higher extent than the other excited states. The metastable 2^3S_1 atoms are polarized by optical pumping with circularly polarized light. The polarized He* atoms thus obtained give rise to polarized free electrons in the discharge. An exit canal in the discharge cell allows extraction of electrons and analysis of their polarization by a Mott detector.

Figure 5.16 explains how the metastable atoms are polarized by optical pumping. By irradiating with infrared circularly polarized σ^+ light of

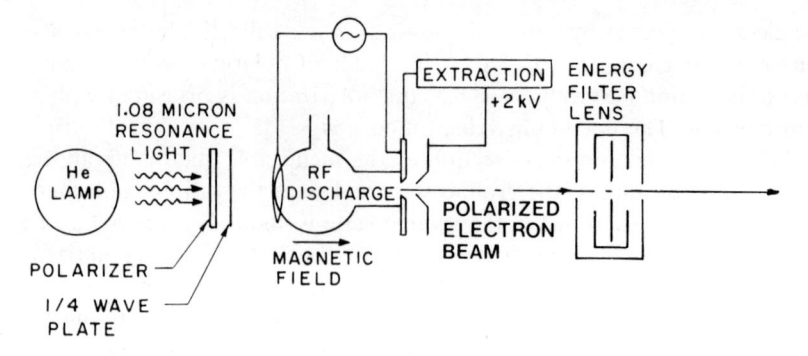

Fig. 5.15. Polarized electrons from an optically pumped He discharge [5.25]

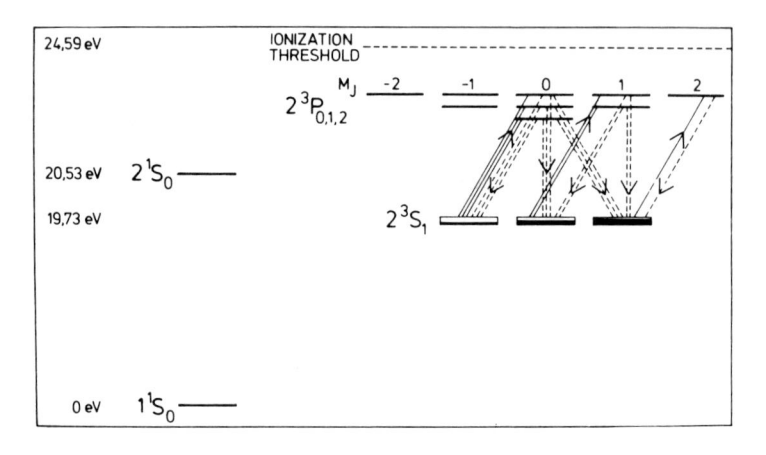

Fig. 5.16. Optical pumping in helium. Not drawn to scale. The $2\,^3P_{0,1,2}$ levels lie so close together that they are all excited by the 1.08 μm line

wavelength 1.08 μm one obtains transitions into the $2\,^3P$ states with the selection rule $\Delta M_J = +1$. The subsequent spontaneous emission is governed by the selection rules $\Delta M_J = 0, \pm 1$, so that only some of these transitions go back to the initial states; the rest go to higher M_J of the state $2\,^3S_1$. By continuous pumping, the population of the state $2\,^3S_1$ ($M_J = +1$) is therefore increased at the expense of the two other sublevels of $2\,^3S_1$. Since in the $2\,^3S_1(M_J = +1)$ state both electron spins are parallel to the incident direction of the light, one obtains a partial polarization of the He*($2\,^3S_1$) atoms in this direction. These metastables are thus labeled by their polarization. If there are polarized free electrons in the discharge they could only have originated from these atoms.

5.5 Collisional Ionization of Polarized Atoms   151

On the other hand, there are numerous processes in the discharge that can produce unpolarized electrons, for example, collisions of ions with the walls of the exit canal and other secondary processes. The fact that, in spite of this, the extracted electron beam was still found to be 10% polarized, shows that a large fraction of the free electrons in the discharge originates from the polarized He* atoms (an estimate showed that—since the metastable atoms are only partially polarized—in the ideal case where all extracted electrons are produced by collisions between metastable atoms, 30–40% polarization should be measured).

This proves that collisional ionization of unpolarized He atoms (which are at least a factor 10^5 more numerous than the polarized He* atoms) by electrons is not the main source of the free electrons. Ionization of the polarized He* atoms by electron collisions (ionization energy 4.7 eV) can, however, also be excluded: Since in the discharge there are not enough electrons available with energies > 4.7 eV (estimated on the basis of the electron temperature of 2.5 eV), this process cannot possibly balance the large electron loss rate due to diffusion to the walls, as would be necessary for maintaining a steady discharge. Exchange scattering of slow electrons by the polarized He* atoms does not release enough polarized electrons either, as another estimate by the authors shows. Thus Penning ionization, that is, ionization by collision of two metastable atoms, remains as the predominant electron production process.

Let us consider this reaction in more detail. When a He*(2^3S_1, $M_J = +1$) atom collides with a He*(2^3S_1, $M_J = 0$) atom, the reaction can be written

$$\text{He*}(2^3S_1) + \text{He*}(2^3S_1) \rightarrow \text{He}(1^1S_0) + \text{He}^+ + e,$$
$$\uparrow\uparrow \qquad \uparrow\downarrow + \downarrow\uparrow \qquad \uparrow\downarrow - \downarrow\uparrow \quad \uparrow \qquad \uparrow$$

where the spin eigenfunctions of the various particles are given symbolically. Conservation of total spin, which is assumed in the above reaction, has been separately proved by the authors [5.26]. They showed that with strong polarization of the metastable atoms, the production rate of free electrons decreases. This has the following explanation: With increasing polarization, the Penning ionization must increasingly occur with metastable atoms in the state $M_J = 1$. Since

$$\uparrow\uparrow + \uparrow\uparrow \rightarrow \uparrow\downarrow - \downarrow\uparrow + \uparrow + \uparrow$$

is incompatible with spin conservation, the decrease of the free electrons with increasing polarization shows that spin conservation holds. Furthermore, the observation of the decrease of the free electrons provides additional evidence for the fact that Penning ionization is the predominant source of ionization in the discharge.

Polarized electrons are also produced by the reaction

$$\text{He}^*(2^1S_0) + \text{He}^*(2^3S_1) \rightarrow \text{He}(1^1S_0) + \text{He}^+ + e$$
$$\uparrow\downarrow - \downarrow\uparrow \qquad\qquad \uparrow\uparrow \qquad\qquad \uparrow\downarrow - \downarrow\uparrow \quad \uparrow \qquad \uparrow$$

in which metastable atoms in the singlet state are involved. The electrons produced by Penning ionization of two H*(2^3S_1) atoms with $M_J = 0$ are, however, unpolarized.

The conclusions drawn here were confirmed by measuring the electron polarization during the afterglow: When the discharge voltage is terminated the electrons rapidly come to thermal equilibrium with the gas atoms so that electron-impact ionization can no longer take place. Nevertheless, for 10^{-3} to 10^{-4} s (during the so-called afterglow) charged particles are still produced. It is known that the Penning ionization is responsible for this. There are no competitive electron-production processes in the afterglow, so that an increase of the polarization is to be expected. Indeed, the polarization measurements on electrons which were extracted during the afterglow yielded polarizations of up to 17%.

These measurements have demonstrated that while collisions between metastable atoms are the only source of free electrons in the afterglow, they also produce most of the electrons in the active discharge. The process discussed here is but one of many (e.g., chemi-ionization [5.27] or atom-surface collisions [5.28]) in which electron polarization can be used as a diagnostic tool for the analysis of collision processes [5.29–31]. Such processes can also be considered with a view to building a source of polarized electrons. We will return to this point in Section 7.3.

6. Polarized Electrons from Solids

6.1 Magnetic Materials

Polarized free electrons can be obtained by photoemission or field emission of the spin-oriented electrons from magnetized materials. Such measurements provide interesting, in some cases striking, information on the structure of magnetic solids. Magnetic insulators and semiconductors tend to give results that can be interpreted within the framework of current models, whereas the results from metals call for a change of either the band theory of magnetism or the existing models of electron emission.

In magnetized solids there are electrons whose spins have a preferential orientation. An obvious idea is to try to somehow extract these polarized electrons and thus to obtain polarized free electrons. This could, for example, be done by the photoeffect or by field emission. However, for a long time all attempts of this kind failed. This was probably due to inadequate vacuum conditions in some of the experiments and to inexpedient electron-optical arrangements in others (when, for example, one tried to extract the polarized electrons perpendicular to the external magnetizing field B, which causes the desired electrons to drift in the direction $E \times B$).

Only in the past few years have polarized electrons been successfully extracted from magnetic materials. Several experiments were made on ferromagnetic metals and insulators and on antiferromagnetic semiconductors. The results are not yet fully understood in every case.

In the classical ferromagnets Fe, Co, and Ni the magnetism is caused by the 3d electrons. For the description of the magnetism in these materials the band model is widely used. In this model, the exchange interaction which is responsible for the ferromagnetism causes an energy shift of the e↑ and e↓ in the 3d band. The shift is determined by the strength of the exchange interaction and has different directions for e↑ and e↓, as shown schematically in Fig. 6.1. Both the spin-up and spin-down bands are filled up to the Fermi energy, which means that there are more e↑ than e↓. For nickel there are 5 electrons per atom in the $3d_\uparrow$ band and 4.4 electrons per atom in the $3d_\downarrow$ band. Accordingly, there is a resulting magnetic moment of $0.6\mu_B$ per atom. Above the Curie temperature, the energy shift caused

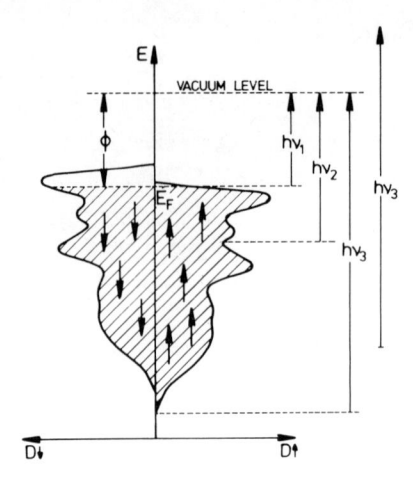

Fig. 6.1. Schematic diagram of the 3d-band splitting below the Curie temperature. E_F = Fermi energy, ϕ = work function, D = density of states. The 3d band is overlapped by a 4s band (not shown here) which contains 0.6 electrons per atom

by the exchange interaction disappears. Both bands then contain equal numbers of electrons; thus the ferromagnetism disappears.

Apart from the ferromagnetic metals, the europium and gadolinium compounds which are ferromagnetic or antiferromagnetic at low temperatures have also been studied. In these compounds (e.g., the europium chalcogenides EuO, EuS, etc.) it is mainly the 4f electrons of the europium or gadolinium atoms that are responsible for the magnetism, as is illustrated by the schematic energy level diagram of Fig. 6.2. There are 7 electrons per europium atom in the localized 4f states. Since $l = 3$ for f states, this means that half of the $2(2l + 1) = 14$ states available are occupied. The Pauli principle still allows all the electrons to have the same

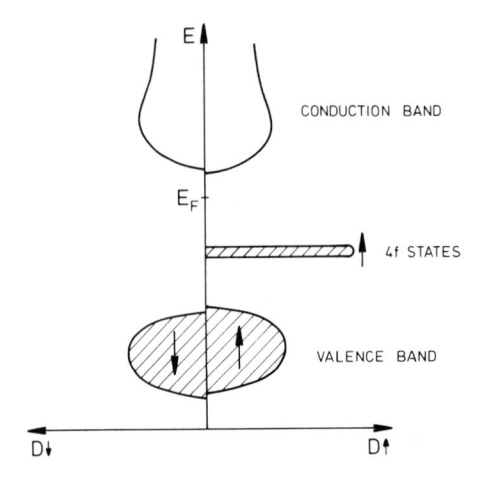

Fig. 6.2. Schematic diagram of the energy states in europium chalcogenides

spin direction. In accordance with Hund's rule, this state, which has the spin quantum number $\frac{7}{2}$, occurs in europium. It is evident that extraction of the electrons from the 4f states should yield highly polarized electrons.

6.1.1 Photoemission

If nickel is irradiated with short-wave light of frequency $v \geq v_1$ (see Fig. 6.1), 3d electrons are photoemitted. With the frequency v_3 the 3d electrons can be knocked out irrespective of their energetic position. If one uses the numbers from page 153 (and ignores complications which have yet to be mentioned) one would expect a polarization $P = (5 - 4.4)/(5 + 4.4) \approx 6\%$ if the sample is magnetically saturated. Since the magnetization, and therewith the preferential orientation of the magnetic moments, is parallel to the magnetizing field **B**, the preferential orientation of the spins is antiparallel to **B** due to the negative gyromagnetic ratio. We therefore mean here by positive polarization that most of the magnetic moments are parallel and most of the spins antiparallel to **B**.

If one irradiates with the frequency v_1, so that only electrons from the immediate region about the Fermi level can be emitted, then according to Fig. 6.1 one would expect more e↓ than e↑ since the density of states $D\!\downarrow$ is larger than $D\!\uparrow$ near the Fermi level. The polarization should therefore be negative.

These examples and Fig. 6.1 show that the polarization of the photoelectrons is determined by the photon energy, the shape of the density-of-states curve $D(E)$, the energy shift of the ↑ and ↓ bands, and the position of the Fermi level. Thus we see that measurements of the photoelectron polarization as a function of the photon wavelength yield detailed information on the electronic structure of magnetic materials. Even more could be learned if, with the use of electron spectrometers, the dependence of the polarization on the photoelectron energy were measured.

The only extensive experimental studies on the polarization of photoelectrons from magnetized materials are being made by a group in Zürich [6.1-7]. Because the results are reproducible only with clean surfaces, the samples are produced in ultrahigh vacuum, either by evaporating in situ or by cleaving single crystals. A strong external field is needed to magnetize the samples in order to reach magnetic saturation. The magnetic field is, like the electric extraction field, perpendicular to the photoemitting surface of the sample so that the extraction of the electrons is not hindered by the Lorentz force (see Fig. 6.3). The photoelectrons released by uv light of variable wavelength are deflected through 90° by a cylindrical condenser so that they leave the region of the light beam and their longitudinal

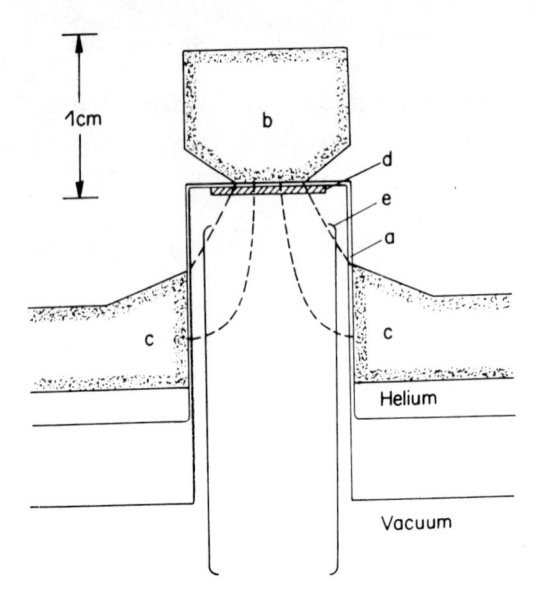

Fig. 6.3. Magnetic photocathode for the emission of polarized electrons [6.3].
a = vacuum envelope, b, c = poles of the electromagnet, d = sample, e = extraction electrode,
---- magnetic field

polarization becomes transverse (see Subsect. 3.6.3). This entire arrangement is placed in a He cryostat because the Curie points of many of the samples are at low temperatures. The polarization is measured with a Mott detector after the electrons have been accelerated to 100 keV.

Examples of experimental results are given in Figs. 6.4 and 6.5. Figure 6.4 shows the dependence of the photoelectron polarization on the strength of the external magnetic field. With nickel, the polarization (which is proportional to the magnetization) clearly reflects the well-known saturation behavior of the magnetization curve. Whereas in the Ni curve the photoelectrons originated from near the Fermi level, with pure or lanthanum-doped EuO crystals the photon energy was chosen so that most of the photoelectrons came from the $4f$ levels. The high polarization which is expected in emission from these levels was actually found, as shown in Fig. 6.4. The fact that in this case no saturation occurs is explained by the influence of the surface layer: The measured photoelectric magnetization curves represent the magnetic behavior of a thin sheet of material at the surface, whose thickness is determined by the mean free path of the photoexcited electrons in the material. The solid curves for EuO and $Eu_{1-x}La_xO$ have been calculated under the assumption that the bulk material saturates at the kinks of the curves, whereas the magnetization of the surface layer increases further as the field increases.

We will not list here all substances on which such measurements have been made (it should be mentioned that with iron $P > 50\%$ was found

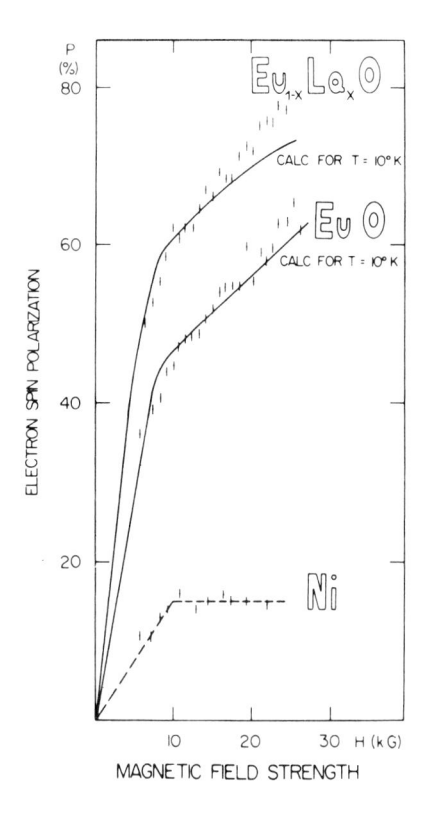

Fig. 6.4. Dependence on the magnetic field strength of the polarization of photoelectrons from the neighborhood of the Fermi level in Ni films and from the 4f levels of EuO and $Eu_{1-x}La_xO$, where $x = 1$ at.% [6.5]

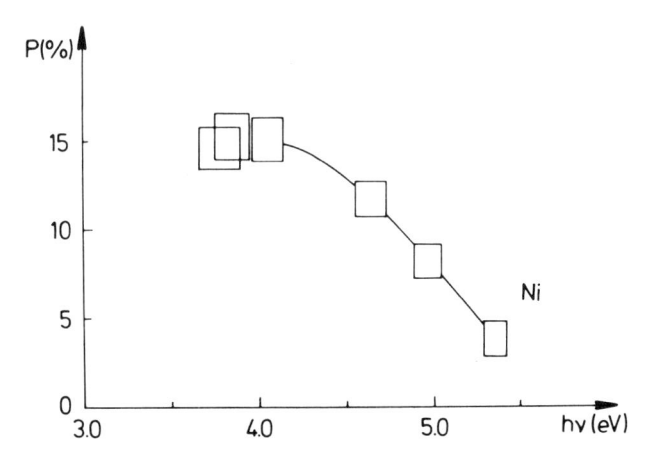

Fig. 6.5. Dependence of the polarization on the photon energy $h\nu$ for Ni films [6.2]. $T \approx 10$ K

[6.6]) but rather discuss the wavelength dependence of the polarization using our sample substance Ni. In the measurement [6.2] shown in Fig. 6.5 the external magnetic field was held constant at a value large enough to produce magnetic saturation. Cesium was evaporated onto the nickel film to reduce the work function and thus reach an experimentally more convenient wavelength range (much less than a monolayer of Cs was deposited so that the polarization would not be appreciably affected). The decrease of the polarization curve with increasing photon energy can be explained as follows: At the threshold, which corresponds to the frequency v_1 in Fig. 6.1, the polarization is determined by the difference between the densities of states $D\uparrow - D\downarrow$ at the Fermi level. As the frequency increases, the difference $D\uparrow - D\downarrow$ and therewith the degree of polarization decreases, since apart from the relatively strongly polarized electrons from the Fermi level the weakly polarized electrons from the deeper regions of the band are then emitted. The exact shape of the polarization curve depends on the details of the band structure.

There are, however, severe discrepancies between the predictions of the band model and the results obtained [6.2, 6]: The polarization of the electrons emitted from the Fermi level is found to be positive. When discussing Fig. 6.1 we saw that the band model predicts a negative polarization of these electrons. Furthermore, density-of-state curves like those shown in Fig. 6.1 should result in a much more pronounced dependence of the polarization on the photon energy than that given by Fig. 6.5.

One must, of course, check whether these discrepancies between theory and experiment might originate from the simplifying assumptions we have made. We have implicitly assumed that the polarization of the photoelectrons is identical to or at least proportional to the polarization in the solid. This assumption would certainly not be valid if—in analogy to the Fano effect—the initial polarization were changed by the photoabsorption process, or if it were highly probable that the spins of the electrons would flip en route to or when emerging from the surface. Indeed, an exact treatment of all these processes is difficult; estimates [6.3] and additional measurements [6.4] have shown, however, that in the experiments made so far, no appreciable errors have been caused by these processes. The fact that a small fraction of the electrons is in the 4s band (in our numerical example 0.6 of the total 10 outer electrons per atom) and may have some polarization there likewise has no appreciable effect. A detailed discussion of the possible error sources, the influence of surface layers, and cross checks with other experimental results can be found in a review article by SIEGMANN [6.1].

It is interesting to note that the results on the spin direction near the Fermi surface agree with those of tunneling experiments using junctions of thin superconducting Al films with ferromagnets in a magnetic field [6.8]. The field acts on the electron spins causing a shift of the energy spectrum of the quasiparticles in the superconducting Al by $\pm\mu B$ for e↑ and e↓, respectively. This energy splitting allows electrons of either spin direction to be selected from the electrons tunneling from the ferromagnetic electrode of the junction and permits tunneling investigations of the electron polarization in magnetic metals. This method has been used for measuring the conductance of metal–Al_2O_3–Al junctions as a function of the applied voltage. At a suitable voltage it is predominantly the up spins that make the tunneling current; when the voltage is reversed it is the down spins. These measurements, which allow studies of the spin polarization of electron states within an energy range of 1 meV below the Fermi level, have been made for junctions with Fe, Co, Ni, and Gd and agree with the results of the photoemission studies discussed before. Since we deal in this book with free polarized electrons we will refrain from further discussion of this tunneling detector for spin polarization in metals.

The spin polarization spectra show that for $3d$ metals either the theoretical interpretation of magnetism or the existing models of electron emission will have to change. However, for the ferromagnetic and antiferromagnetic Eu and Gd compounds mentioned above, the polarization spectra do agree with the theoretical predictions. According to Fig. 6.2, one expects valence states, $4f$ states, and conduction states to emit electrons of different polarization. The contribution of the $4f$ states should be particularly significant since they contain only electrons of one spin direction. This was confirmed by measurements of the photoelectron polarization as a function of photon energy [6.1]. The sign of the polarization and its dependence on the photon energy was as expected or could be easily interpreted within the framework of current models. The same is true of results obtained from ferrites [6.9].

An interesting experimental feature of electron-spin spectroscopy results from the fact that the photoelectric probing depth is determined by the small mean free path of the photoelectrons. The method is therefore appropriate for studying surface magnetism or for measuring magnetization in very thin films. Assuming a light spot of 1 mm^2 and a mean free path of the order of 10 Å, we see that the volume of material required is $\approx 10^{-9}$ cm^3. A sample of that size would not give a detectable signal in a conventional magnetometer.

6.1.2 Field Emission

Another possibility for extracting electrons from solids is field emission in a high electric field. If the field strength is E near the surface of the sample,

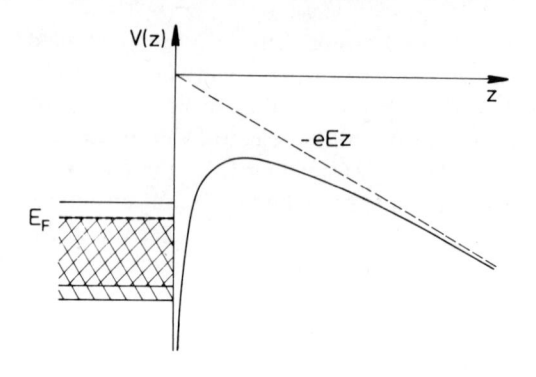

Fig. 6.6. Model for the field emission from a ferromagnetic metal

the potential energy of an electron outside the surface is roughly given by $-eEz$. In this expression the electrostatic charges induced in the sample by the electron are not yet taken into account. They reduce the potential energy so that the potential has the shape of the solid curve in Fig. 6.6. By tunneling through this potential barrier, the electrons of the sample can reach the free states. The tunneling probability decreases sharply with increasing width and height of the potential barrier so that virtually only states within a few tens of millivolts below the Fermi level contribute to the field emission.

Figure 6.6 shows schematically the case of a ferromagnetic metal in which the valence band is believed to be split due to exchange interaction. For the 3d band of nickel, the splitting is illustrated in Fig. 6.1 where the superposition of the 4s band mentioned in the figure caption must also be taken into account.

For the polarization of the field-emitted electrons one might expect approximately $P = (N_\uparrow - N_\downarrow)/(N_\uparrow + N_\downarrow)$, where $N_\uparrow$ and $N_\downarrow$ are the numbers of $e\uparrow$ and $e\downarrow$ in the neighborhood of the Fermi level. The extent to which $N_\uparrow$ and $N_\downarrow$ differ from each other depends on the density-of-states curves near the Fermi level.

Here again the assumption has been made that the polarization of the emitted electrons is the same as the polarization within the ferromagnetic metal. One cannot, however, assume this in general: Not only does the exchange interaction shift the position of the energy bands for $e\uparrow$ and $e\downarrow$, but it also causes somewhat differing shapes of the potential barrier for each of the two spin directions [6.10]. Thus the $e\uparrow$ and $e\downarrow$ will have different tunneling probabilities so that the polarization measured externally need not be the same as the internal polarization. The fact that a small fraction of the electrons near the Fermi level is in the 4s band has similar consequences. The tunneling probability of the 4s electrons which have only slight, if any, polarization is considerably larger than that of the 3d electrons [6.11]. The polarization of the field-emitted electrons is therefore strongly diminished and is by no means the same as the polarization within the solid. The situation is here quite different from that in photoemission. The photoelectrons, due to their higher energy, have a much greater probability of leaving the metal so that the relative differences in escape probabilities for the different states are small and affect the observed polarization only slightly.

Measurements that have been carried out on ferromagnetic metals have mainly concentrated on nickel and gadolinium and yielded polarizations in the 10% region [6.12, 13]. The polarization found for nickel was negative. Thus one finds here—contrary to photoemission—the sign that is expected from the band model. However, with field emission, due to the aforementioned difficulties in relating the observed polarization to the polarization within the material, a quantitative comparison of experimental results with theoretical calculations of the polarization at the Fermi level is rather problematic. Another problem is that large fluctuations of the polarization as a function of the magnetic field strength have been observed. Further independent measurements are desirable in order to understand all the details of this interesting experiment before definite conclusions can be drawn from the results.

An example of an experimental setup for measuring polarization by field emission [6.14] is shown in Fig. 6.7. The investigations were made on ferromagnetic EuS, a thin film of which was evaporated in situ onto a

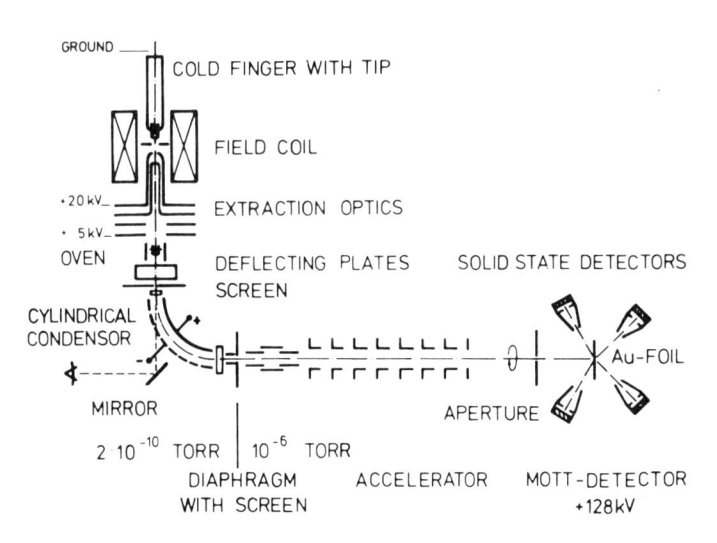

Fig. 6.7. Arrangement for measuring the polarization of electrons field emitted from ferromagnetic materials [6.14]

tungsten tip. The tip was mounted on a cold finger which could be cooled down to 14 K (Curie point of well-annealed films is 16.6 K). A field of up to 20 kG was used to reach magnetic saturation. This high field was produced in pulses of 0.25 ms duration with a repetition rate of 6–30 pulses per minute; the field coils were cooled with liquid nitrogen. The electric field strength of more than 10^7 V/cm required for emission from the tip was produced by a voltage of 20 kV. The arrangement was placed in ultrahigh vacuum ($3 \cdot 10^{-10}$ Torr) to avoid contamination of the emitting material. The longitudinal polarization of the field-emitted electrons was transformed into transverse polarization by a cylindrical condensor. The electrons then passed through a diaphragm from the ultrahigh vacuum into an accelerator tube leading to the Mott detector where the polarization was determined.

The measured degree of polarization depended strongly on the film thickness, which was considerably influenced by the annealing temperature (annealing was necessary to produce an ordered structure in the EuS film). The values measured under optimum conditions were $(89 \pm 7)\%$.

The experimental results can be explained as follows: The high polarization of 89% is caused by the europium $4f$ states shown in Fig. 6.2. If there is a sufficiently strong field at the tip—i.e., if the potential barrier is narrow enough (see Fig. 6.8)—the totally polarized electrons in the $4f$ states can tunnel into the vacuum and electrons of the tungsten substrate can reach the vacuum by tunneling over the $4f$ states. Since europium sulfide is a good insulator at low temperatures, the external electric field penetrates into the EuS film and causes a drop $-(1/\varepsilon)eEx$ of the potential energy (ε = dielectric constant of EuS). Figure 6.8 shows the correspond-

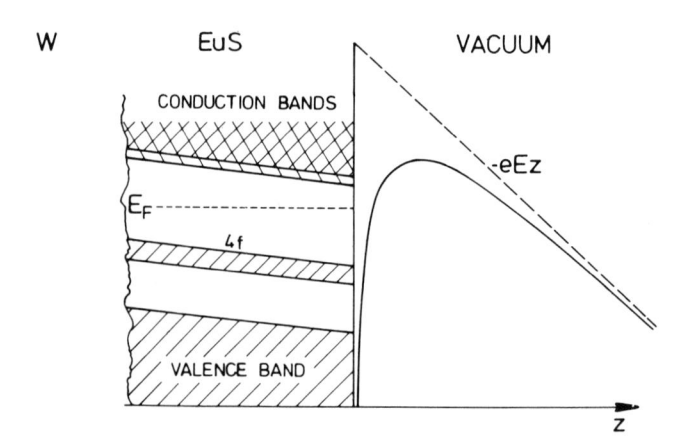

Fig. 6.8. Field emission from EuS

ing slope of the energy bands. With sufficiently thick films and strong enough fields, EuS conduction states can sink even below the Fermi level E_F, which has the same height in the W tip and the EuS coating (the matching of the Fermi levels is caused by a potential drop in the boundary layer). Electrons from the tungsten substrate can then tunnel into the EuS conduction band and from there into the vacuum. Since the conduction band is split due to the interaction with the 4f states, the EuS tunnel barrier is different for different spin directions. The polarization found under these conditions was 20–50% depending on the temperature. With very thin EuS films of ≈ 1 layer no polarization could be detected. This can be understood by assuming that these films are nonmagnetic.

We may interpret the experiment by saying that EuS has been used as a polarization filter for the electrons emitted by the W tip into the vacuum. The physical mechanism of this spin filter is based on the spin-dependent interaction of those electrons with the 4f states. Although the discussion of Section 1.2 showed that "conventional" spin filters do not work with electrons, we see now that "unconventional" spin filters for electrons do exist. Discussions of the kind presented in Section 1.2 should not discourage us from searching for electron polarization filters which are just as effective as those for light and for atoms!

In the spin filter discussed here the polarization varies with the thickness and structure of the EuS film, the temperature, and the strength of the magnetic field. The high polarization of 89% can therefore be obtained only if the optimum conditions are strictly maintained.

6.2 Nonmagnetic Materials

By using circularly polarized light, one can obtain polarized photoelectrons also from nonmagnetic solids. The experimental results for alkali metals can be attributed to the energy-band splitting caused by spin-orbit interaction. This is confirmed by experiments made with GaAs crystals for which the polarization curve obtained can be easily explained by means of the well-known band structure. Studies of the polarization in photoemission from nonmagnetic solids allow one to determine the extent of the energy-band splitting, a quantity which is presently unknown for many materials.

As we saw in Chapter 5, one does not need materials with oriented spins in order to produce polarized photoelectrons. Photoionization of unpolarized alkali atoms with circularly polarized light yields, for example, polarized electrons, too.

An obvious question is, whether this works only with free atoms or whether solids could also be used. The first experimental answer to this question was affirmative [6.15]. The measurements were carried out with the apparatus shown in Fig. 5.5, the only major change being that the target was no longer an atomic beam. Instead, the alkali atoms were evaporated on a substrate in normal high vacuum at a rate of about 50 atomic layers per second.

Comparison of the measured curve in Fig. 6.9 with that for free cesium atoms (Fig. 5.6) shows some characteristic differences: The polarization maximum is shifted to longer wavelengths and no longer lies in the uv, but rather in the easily accessible visible wavelength region. The degree of polarization is much smaller, but the photocurrents obtained from solid samples are about 1000 times larger than those from atomic beams, which are targets of much lower density.

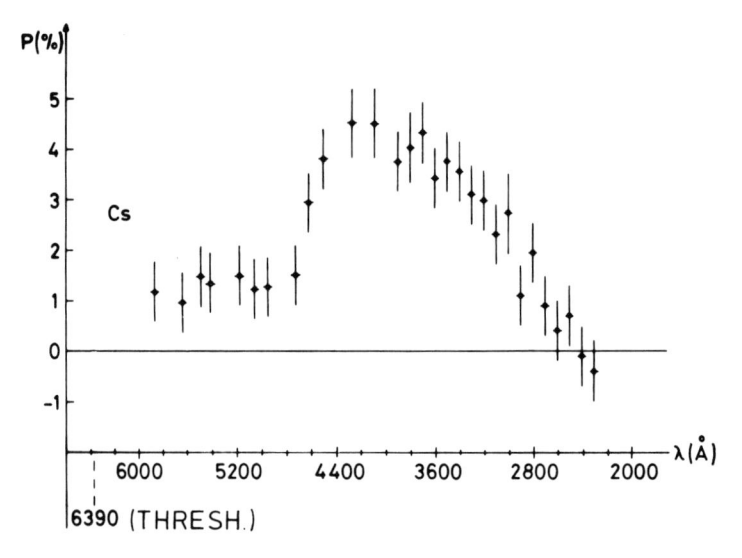

Fig. 6.9. Polarization of the photoelectrons from solid cesium produced by circularly polarized light [6.15]

Films of other solid alkalis also yielded polarized electrons when photoionized with circularly polarized light, as shown in Fig. 6.10. It can be seen that the polarization decreases with decreasing atomic number. This is to be expected if, as in Section 5.2, spin-orbit coupling is responsible for the polarization. Within the error limits indicated in Fig. 6.10, no polarization was found for Na and Li with the aforementioned experimental arrangement.

What has already been said in other sections of this chapter applies to an even greater extent to the results now being discussed: The polarization

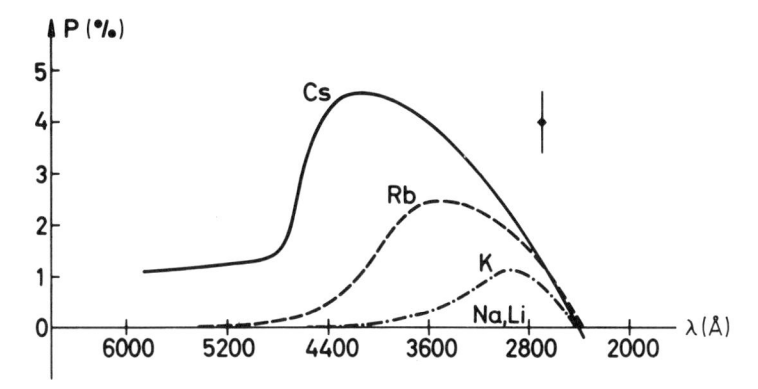

Fig. 6.10. Comparison of the photoelectron polarization for different alkali films. The error bar shown applies to all three curves [6.15]

phenomena obtained with solids are less well understood, quantitatively, than those obtained with free atoms. Until now, the above results have been explained only in principle. A model calculation has shown that the spin polarization of the photoelectrons can be explained by the energy-band splitting in solids caused by spin-orbit interaction [6.16].

The energy-band splitting, which is analogous to the spin-orbit splitting of the energy levels in free atoms, is schematically illustrated in Fig. 6.11. The left-hand side represents two energy states for a certain magnitude and direction of the wave vector k. The lower state Δ_1 belongs to the valence band,[1] whereas the upper state Δ_5 lies above the vacuum level E_∞. Elec-

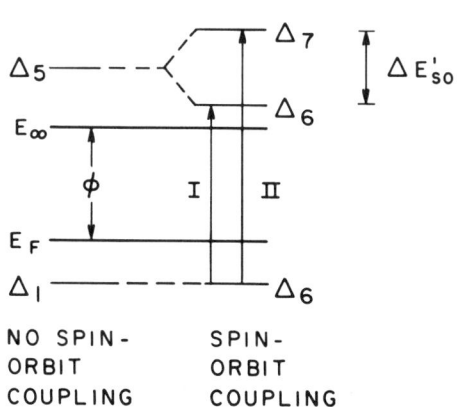

Fig. 6.11. Allowed transitions between Δ states induced by circularly polarized light. $\Delta E'_{so}$ = spin-orbit splitting, E_F = Fermi energy, E_∞ = vacuum level, ϕ = work function

---

[1] The subscripts of the band symbols are related to the crystal symmetry and need not be explained for the qualitative description given here. For an introduction to the connection between crystal symmetry and band structure see [6.17].

trons with energies $E \geq E_\infty$ can escape from the solid into the vacuum. When the magnitude of our specific wave vector is changed, the position of each energy state is shifted within the band of the allowed energies. (Δ band for the specific example chosen, where k is directed to a certain corner of the Brillouin zone. For other directions of k the energy bands are labeled differently).

The splitting of the energy states due to spin-orbit interaction is shown on the right-hand side of Fig. 6.11 for a certain k. Variation of the magnitude of k yields the split Δ band. The figure takes into account the fact that in the case discussed it is only the upper band that splits. We now consider direct interband transitions from states below the Fermi level E_F to states above the vacuum level E_∞ and look into the polarization of the photoelectrons thus produced.

Since neither quantitative calculations of the band splitting in Cs nor complete wave functions are available, it is impossible at the moment to calculate the curve $P(\lambda)$ quantitatively. The salient features of the polarization curve can, however, be predicted, knowing only the spin and angular parts of the wave functions. We have already seen this in Chapter 5. Spin and angular parts of a wave function are determined by the symmetry properties of the system considered and can therefore be derived with the help of group theory. This is much less complicated than the numerical calculation of the radial part. Evaluation of the matrix elements with the basic functions which contain the angular and spin parts and which are analogous to the coupled wave functions used in Sections 5.2 to 5.4 yields the selection rules for the various interband transitions.

In transition I of Fig. 6.11, for example, the matrix element for excitation of the spin-up state disappears if σ^+ light is used. This means that in this transition totally polarized e↓ are produced. In transition II the matrix element for excitation of the spin-down contribution disappears, so that only e↑ are ejected. Transitions between energy states associated with other directions of k can be treated similarly. It frequently occurs, however, that the matrix elements for both the ↑ and ↓ parts of the final state are different from zero. In this case a quantitative calculation of the resulting polarization requires knowledge also of the radial parts of the wave functions.

The discussion indicates that there are further appealing aspects of this polarization phenomenon besides the production of polarized electrons: The photoelectrons produced by transitions I and II have not only different polarizations but also different kinetic energies, the energy difference being given by the spin-orbit splitting $\Delta E'_{SO}$ of Δ_5. Accordingly, measurement of the polarization as a function of the photoelectron energy E_{kin} should be of considerable help in the analysis of the spin-orbit splitting of energy bands.

In the aforementioned experiment all the photoelectrons emitted from a polycrystalline Cs target were detected independent of their energy. This means that the observed polarization was the average over the photoelectron energy and over the transitions between the energy bands associated with the various directions of k. Accordingly, the observed polarization was very small. For investigations of the spin-orbit splitting of energy bands the polarization of photoelectrons emitted from oriented single crystals within a small solid angle should be measured as a function of the photoelectron energy E_{kin} and the photon wavelength λ. In special cases, by suitable combination of E_{kin} and λ, it is possible to select specific interband transitions also in polycrystals.

If we apply our considerations to materials with better known electronic structure, it is possible to calculate the photoelectron polarization quantitatively [6.18, 19]. Contrary to the case of cesium, the spin-orbit splitting of the energy bands in GaAs is well known [6.20]. Relevant $E(k)$ curves for the specific k direction generally denoted by Λ are shown in Fig. 6.12. Evaluation of the matrix elements shows that transition I from the valence band maximum to the conduction band minimum yields a polarization of -0.5 if σ^+ light is used for excitation.[2] Similarly, for transitions II, III, IV one obtains $P = +1, -1, +1$, respectively. Accordingly, the photoelectron polarization as a function of the photon energy should have a pronounced structure.

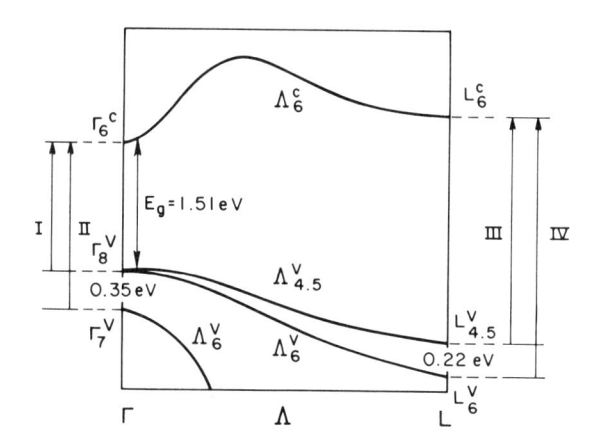

Fig. 6.12. Energy bands of GaAs (not to scale) [6.20] and transitions induced by circularly polarized light. The superscripts v and c denote conduction and valence bands

---

[2] As in Sections 5.2 to 5.4, the polarization of the photoelectrons is positive if their spins are preferentially oriented in the direction of light propagation (=direction of photon spins for σ^+ light).

This has been confirmed by an experiment [6.19, 21] using the ultrahigh-vacuum apparatus described in Subsection 6.1.1. Needless to say, the magnetizing field was not needed in the present experiment. Normally, electrons with excitation energies in the range shown in Fig. 6.12 cannot escape from the GaAs sample. Instead, they stay in the conduction band until they recombine. As a matter of fact, the polarization of conduction electrons excited by circularly polarized light has been analyzed in earlier experiments by measuring the polarization of the luminescence light and by resonance techniques. In the present experiment, all the electrons that were excited to the conduction band could escape into the vacuum since the vacuum level E_∞ of the GaAs crystal had been lowered by applying alternating layers of cesium and oxygen to the surface of the sample. Measurement of the polarization with a Mott detector resulted in the curve of Fig. 6.13.

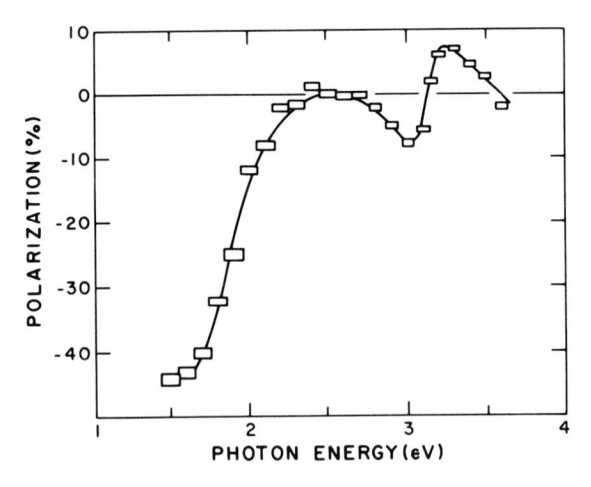

Fig. 6.13. Polarization of photoelectrons obtained from GaAs + CsOCs at $T \lesssim 10$ K with σ^+ light [6.19]

The measured value $P = -45\%$ near threshold is close to the predicted value of -50%. As the photon energy is increased above 1.86 eV, a strong admixture of $e\uparrow$ due to transition II occurs. With a further enhancement of energy one finds increased cancellations of the positive and negative polarizations produced by the various possible transitions, until transition III gives rise to a sudden increase in the number of $e\downarrow$. As the photon energy goes up by another 0.2 eV, a positive polarization peak occurs due to transition IV.

The agreement of the experimental results with the theoretical prediction shows that no severe changes of the electron polarization occur by the transport to and through the surface. The trend of the polarization to decrease with increasing photon energy is typical for integral experiments in which the polarization of all the photoelectrons produced by a certain hv is measured. Use of differential electron spectrometers would make possible measurements of $P(E_{kin})$. They are much easier to perform in experiments with nonmagnetic materials than in the experiments discussed in the previous section because no disturbing magnetic fields need be applied here. Such measurements would allow much better separation of the various transitions we have discussed for cesium and GaAs. The high polarization peaks produced by transitions II, III, and IV in Fig. 6.12, for example, could then actually be isolated. Experiments of this kind with materials less well known than GaAs would help considerably in resolving problems like the spin-orbit splitting of energy bands and would improve our knowledge of wave functions in solids.

We conclude this section with the remark that spin polarization has also been reported for field-emission electrons from tungsten [6.22]. Polarizations of up to 13% were found when monocrystalline tungsten tips were placed in strong magnetic fields of up to 25 kG at 80 K. The preliminary hypotheses as to the cause of this polarization effect, which is still somewhat controversial [6.14], are based on the spin-orbit splitting of the partially occupied 5d band in tungsten.

7. Further Applications and Prospects

7.1 Investigations of the Structure of Matter

By diffracting slow electrons on crystals, one obtains polarized Bragg reflections if spin-orbit coupling or—in the case of magnetic materials—exchange interaction plays a role. The analysis of electron-molecule scattering is facilitated when electrons which are scattered by certain atoms of a molecule are labeled by their polarization. In electron microscopy, electron polarization has no significance at present. Irradiation of organic molecules with polarized electrons seems to give a clue as to the origin of the one-handedness of nature. Polarized-electron experiments would provide information in high-energy physics that cannot be obtained from cross-section measurements alone.

In the preceding chapters we have repeatedly referred to applications of spin-polarization phenomena. We have seen that they open up possibilities for investigating the exchange scattering of electrons, the influence of spin-orbit coupling on photoionization and on electron scattering, or for studying the details of certain atomic collision processes. From electron-polarization measurements one has also obtained novel information on the electronic structure of magnetic materials. Whereas in the previous chapters the emphasis was put on the conceptual basis of the polarization phenomena, we shall now discuss in more detail a few additional applications.

7.1.1 Low-Energy Electron Diffraction (LEED)

In Chapters 3 and 4 we dealt with electron scattering from single atoms. If there is a coherent superposition of the scattered electron waves originating from the individual atoms of the target, the distribution of the scattering intensity becomes quite different: It is no longer determined by the individual atoms alone but is considerably influenced by a factor which depends on the structure of the target. This is the basis of the method of electron diffraction for investigating the structure of crystal lattices or molecules.

Let us denote the scattering amplitudes f and g given in (3.51) and (3.52) as f_j and g_j when they originate from the jth atom. For the wave scattered from the total target into the direction θ, we then obtain by in-phase superposition (i.e., by taking the path differences into account) the amplitudes

$$F(\theta) = \sum_j f_j e^{is \cdot r_j} \quad \text{and} \quad G(\theta) = \sum_j g_j e^{is \cdot r_j}, \tag{7.1}$$

where $s = k - k'$ (if k and k' are the wave vectors of the incident and the scattered beam) and r_j is the position vector of the j-th atom.

From these relations, which are derived from simple geometrical considerations in elementary diffraction theory, it follows that the total scattering intensity from the target is proportional to

$$I(\theta) = |F|^2 + |G|^2 = \sum_{j,k} (f_j f_k^* + g_j g_k^*) e^{is \cdot r_{jk}}, \tag{7.2}$$

where $r_{jk} = r_j - r_k$. Here the incident beam was taken to be unpolarized, i.e., the cross section corresponding to (3.65) was used. According to (3.56) and (3.73), the polarization of the scattered beam is determined by the expression

$$i\frac{FG^* - F^*G}{|F|^2 + |G|^2} = i\frac{\sum_{j,k}(f_j g_k^* - f_k^* g_j) e^{is \cdot r_{jk}}}{\sum_{j,k}(f_j f_k^* + g_j g_k^*) e^{is \cdot r_{jk}}}, \tag{7.3}$$

which, according to (3.70), also determines the left-right scattering asymmetry, if one has a polarized incident beam.

If the target is composed of identical atoms, (7.2) can be simplified to

$$I(\theta) = (|f|^2 + |g|^2) \sum_{j,k} e^{is \cdot r_{jk}}. \tag{7.4}$$

In a crystal lattice one has appreciable scattering intensities only in those directions for which the Bragg condition

$$s = k - k' = 2\pi h$$

is fulfilled (h is the reciprocal lattice vector). In other directions one has destructive interference. This is a result of the lattice-dependent factor $\sum_{j,k} \exp(is \cdot r_{jk})$ which appears in (7.4) together with the intensity distribution $|f|^2 + |g|^2$ caused by the individual atoms.

For the polarization, we obtain from (7.3)

$$P = i\frac{fg^* - f^*g}{|f|^2 + |g|^2} \hat{n} \tag{7.5}$$

if we have identical atoms. P depends, in this case, only on the scattering atoms and not on their geometrical arrangement—at least in the approximation made here.

According to Subsection 3.5.2, high values of the polarization (7.5) are found at certain angles. If one selects a diffraction maximum that occurs at such an angle, one has combined the high intensity of the diffraction peak with the high polarization of the scattered beam. In doing this the magic rule that high values of polarization are always associated with low scattering intensities (see Subsect. 3.4.2) will be broken. It is true that in the scattering of fast electrons, due to the rapid decrease of the cross section with increasing angle, most of the intensity goes into the diffraction maxima at small angles, where $P(\theta)$ is very low; with low-energy electron diffraction (LEED), however, high intensity maxima can also be obtained at large angles where there are high polarization values.

So far, our discussion has given only a rough idea of the polarization effects arising in LEED. A rigorous theoretical treatment is difficult, because there are other effects besides the Bragg condition that determine both the intensity and the polarization: Multiple scattering, inelastic processes, and the surface potential barrier (that is, the detailed shape of the transition from the inner potential level to zero potential in the vacuum) play a crucial role [7.1, 2]. Polarizations and intensities in LEED may therefore serve as sensitive probes of these effects.

Studies of spin polarization in LEED are being made in several laboratories. The first experimental result has been obtained with tungsten single crystals [7.3]. It may be of historical interest to note that early measurements of DAVISSON and GERMER [7.4] with nickel crystals can be reinterpreted as a detection of spin polarization in LEED, a fact which was not realized by the authors [7.5].

In the recent experiment mentioned, low-energy electron diffraction from a clean tungsten (001) surface in ultrahigh vacuum has been observed. A narrow slit was cut in the phosphor screen of a conventional LEED apparatus to allow the extraction of a single diffracted beam. The tungsten crystal was mounted on a manipulator which, in conjunction with the movable electron gun and LEED optics, enabled several of the diffracted spots to be directed through the slit and into the fixed entrance aperture of the accelerating column. After acceleration to 100 keV the electrons entered a second ultrahigh-vacuum chamber where their polarization was determined by Mott scattering. For attaining reproducible results, considerable care had to be exercised in the preparation of a clean tungsten surface. This was achieved by repeated heating in oxygen at 10^{-7} Torr with subsequent flashing to 2000 °C. The crystal had to be flashed before each measurement, since the polarization degraded slowly after flashing, decreasing to zero over a period of several hours.

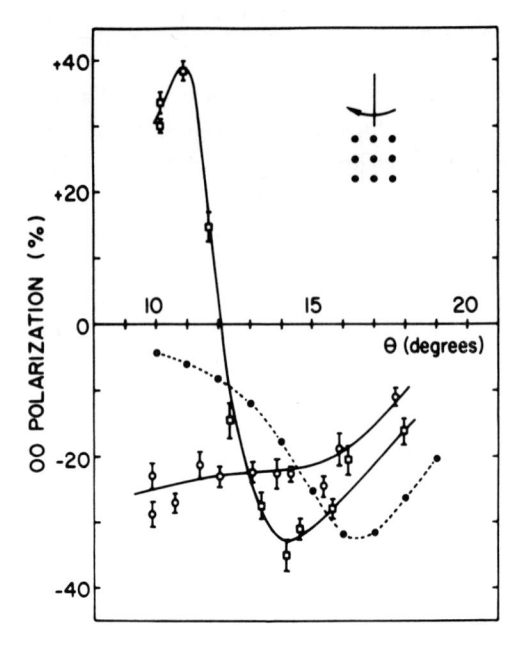

Fig. 7.1. Polarization of the 00 beam as a function of angle of incidence θ for two values of the incident electron energy: ○, 69 eV; and □, 82 eV [7.3]. Dotted line: Calculated polarization for scattering of 100-eV electrons from free tungsten atoms

The measurements were made at electron energies between 45 and 190 eV and at incident angles θ [1] between 10° and 20°. The polarization as a function of energy and diffraction angle has the same pronounced structure as found in scattering from single atoms. Figure 7.1 shows as an example the polarization of the specularly reflected beam for incident angles θ between 10° and 18° (corresponding to scattering angles between 160° and 144°) at two different energies. In comparing these results with the polarizations for scattering from free tungsten atoms, one has to include in the former an inner-potential correction of approximately 15 eV. Thus the 82-eV LEED data should be compared to the results for scattering from free atoms at about 100 eV. Figure 7.1 shows that there is little correlation between these two results. A quantitative explanation of the experimental data has not yet been given.

It is useful to summarize the similarities and differences between the polarization phenomena in electron-atom scattering, treated in Chapter 3, and in LEED. In both cases the underlying physical mechanism is the same. The polarization is caused by spin-dependent scattering. The cross sections for the scattering of the e↑ and e↓ in the unpolarized primary beam differ slightly from each other, which results in a polarization of the

---

[1] Referred to the normal of the crystal surface. $\theta = 0°$ means normal incidence.

scattered beam. Furthermore, one obtains in both cases a left-right scattering asymmetry if one has a polarized incident beam. Accordingly, LEED can be used as a detector of electron polarization in the same way the conventional Mott detector is used. It would be an appropriate spin analyzer for low-energy electron diffraction experiments in which one crystal surface might be used as a polarizer and a second surface as an analyzer. For the many experiments that have to be made in normal vacuum (experiments with gaseous beams, for example), the conventional Mott detector is, however, preferable, since LEED works only with an extremely clean surface. Finally we have the similarity that both in electron-atom scattering and in LEED there is a rotation of the polarization vector of an initially polarized beam [cf. (3.77)]. Neither theoretical nor experimental studies of this effect in LEED have yet been made.

Although the polarization phenomena are, in principle, the same in the two areas compared, there are significant quantitative differences due to the additional influence of the aforementioned solid-state effects that determine the details of the LEED pattern. Consequently, the polarization curves in LEED cannot be expected to be similar to those found in scattering from free atoms. In fact, this is what makes polarization studies in LEED appealing: They provide information on the shape of the surface potential and on the multiple and inelastic scattering processes occurring at and near the crystal surface. This information is not obtainable from studies of the intensity distribution alone; thus these investigations add a new dimension to the study of surfaces.

In this section we have so far discussed only processes in which the polarization is caused by spin-orbit coupling (see Subsect. 3.4.2). If the scattering takes place on magnetized materials, the polarization can also arise from exchange scattering, analogously to the processes discussed in Section 4.1. We shall explain this with an example calculated by Feder [7.6].

Slow electrons are diffracted from a ferromagnetic iron surface. The electron energy is taken to be less than 50 eV so that exchange effects are still appreciable. Only elastically reflected electrons are considered. Exchange processes between electrons with opposing spin directions are thus excluded because every change of the spin state of the target would mean an excitation out of its ground state (excitation of magnons). Accordingly, polarization due to exchange of electrons with different spin directions is not considered.

Nevertheless, the scattered electrons are polarized because the cross sections for e↑ and e↓ differ owing to the exchange interaction. In order to find the scattering cross sections in this many-particle problem it is expedient to start from a statistical description of the target electrons. This yields, besides the electrostatic potential, an exchange potential that is proportional to $\rho^{1/3}$, where ρ is the electron density [7.7]. The densities of spin-up and spin-down electrons in the target are different because we are dealing with a magnetized material. Since, according to the assumption of elastic scattering made above, exchange processes take place only between electrons with the same spin direction, the exchange scattering of e↑ and e↓ is, respectively, caused by the electron densities $\rho_\uparrow$ and $\rho_\downarrow$ in the exchange term of the scattering potential. The different scattering potentials thus arising for the e↑ and e↓ lead to different cross sections for the two halves of the unpolarized incident beam and hence to a polarization of the scattered beam.

Just as in the LEED experiment discussed above the diffraction maxima contain different numbers of e↑ and e↓, but the mechanism causing the polarization is different. By suitable choice of diffraction peaks and electron energies it is again possible to simultaneously achieve polarization maxima and intensity maxima.

If we consider an incident beam normal to the (001) surface of iron, then, for example, the polarization of the 00 beam (i.e., the specularly reflected beam which is here normal to the surface) is as shown in Fig. 7.2. The calculation of the intensity shows that the polarization maxima at 13 and 19 eV correspond to intensity maxima. These results are presumably still far from being realistic since several important factors like inelastic scattering or multiple scattering have not been taken into account. They do, however, represent the first step in a direction which appears to be worthy of further pursuit.

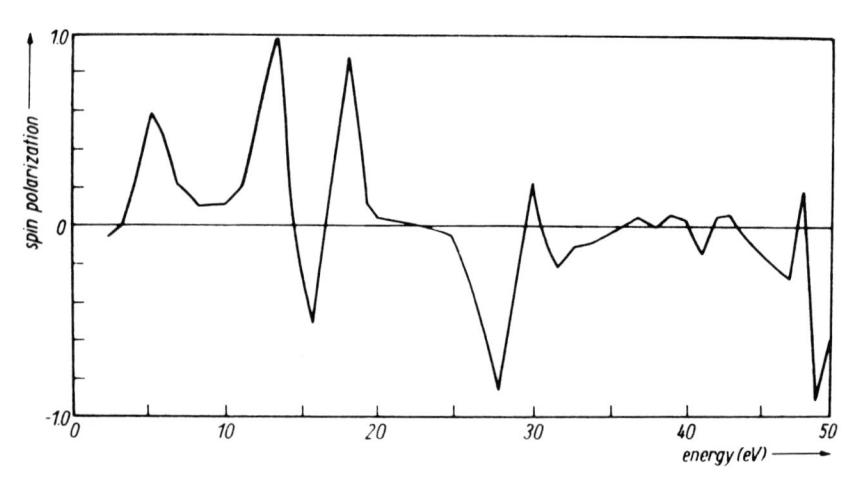

Fig. 7.2. Spin polarization of the 00 beam with the incident beam perpendicular to the (001) surface of a ferromagnetic iron crystal [7.6]

The spin dependence of electron-exchange scattering can be used for studying the structure of magnetic materials. This is analogous to spin-dependent neutron scattering, which led to important discoveries on the structure of magnetic materials. There is, however, an essential difference between the two methods: In neutron scattering the cross sections are very small; they are determined by the short-range nuclear forces and the relatively small interaction of the neutron dipoles with the electron dipoles of the magnetic material. The neutrons can therefore cover great distances in dense material. In electron scattering the cross sections are determined by Coulomb and exchange interactions. Since exchange interaction, which makes the magnetic investigations possible, is significant only at low energies, one must work with slow electrons. The cross sections are then so large that the electrons do not penetrate deep into the material. Electrons are therefore suitable for investigations of the magnetic surface structure, for which neutrons are not appropriate due to their large penetration depths. Accordingly, electron and neutron experiments complement each other in this respect.

This application of spin-dependent electron scattering was first recognized and utilized a few years ago [7.8, 9]. Positive results were reported for LEED experiments on NiO. The structure of this antiferromagnet is shown in Fig. 7.3. It can be seen that the lattice constant of the magnetic

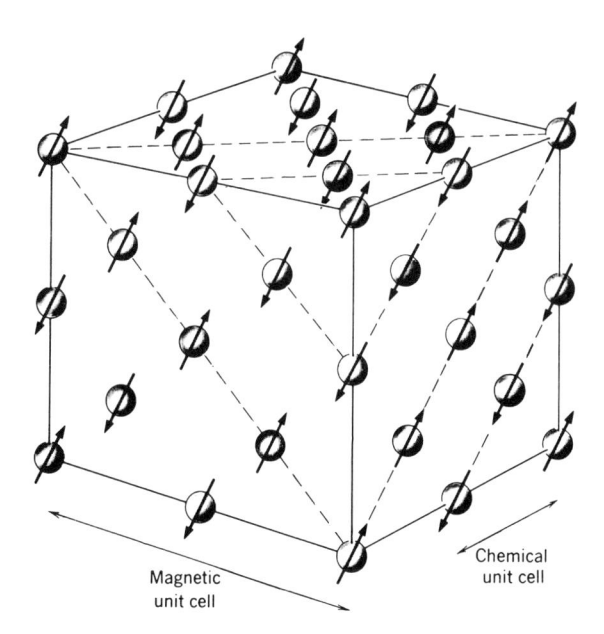

Fig. 7.3. Arrangement of the spins of Ni^{++} ions in NiO. The O^{--} ions are not shown

178   7. Further Applications and Prospects

unit cell is twice as large as that of the chemical unit cell. The Coulomb contribution to the scattering is not influenced by the magnetic properties and thus reflects the structure of the chemical unit cell. The spin-dependent exchange scattering is determined by the magnetic unit cell and should therefore yield additional diffraction maxima half way between those stemming from the chemical unit cell. Observation of these half-order maxima provides a direct means of studying the role of exchange in electron scattering.

Such half-order maxima were indeed observed [7.8, 9] if the temperature of the crystal was kept below the Néel temperature, below which the antiferromagnetism occurs. These investigations on antiferromagnets lie outside the scope of our general topic: They do not have to be carried out with polarized primary beams and the diffraction maxima are not polarized —contrary to what has occasionally been claimed in the literature (see Problem 7.1).

The spin orientation of the electrons in magnetic surfaces can also be determined by studying the polarization of heavy particles (e.g., deuterons) which have captured electrons by interaction with the surface [7.10]. Since the subject of our discussion is polarized free electrons we shall not go into the details of this possibility either.

Problem 7.1: An unpolarized beam of slow electrons impinges on an antiferromagnetic crystal (Fig. 7.3). Is the resulting diffraction pattern determined by the magnetic unit cell (i.e., can one expect to find half-order maxima) or by the chemical unit cell?

Solution: It can immediately be seen that with a polarized beam one would get half-order maxima. This is because the scattering amplitude coming from a particular atom is—as indicated schematically in Fig. 7.4—generally different according to whether the spins of the electron and atom are parallel or antiparallel to each other (if we had free alkali atoms, then according to Section 4.1 the intensities would be $|f|^2 + |g|^2$ in the cases characterized by the solid lines and $|f - g|^2$ for the dashed lines). Accordingly, there can never be complete destructive interference of the bundles coming from two neighboring atoms since their amplitudes are not equal. The lattice constant determining the interference pattern is therefore twice the interatomic distance. If the incident beam consists of $e\downarrow$ the scattered waves indicated by the solid and dashed lines in Fig. 7.4

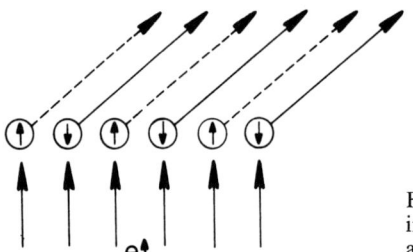

Fig. 7.4. Schematic diagram of the scattering of a polarized electron beam by an antiferromagnet

must be interchanged. Everything else remains the same; one obtains the same diffraction maxima as before.

For an unpolarized incident beam, which consists of e↑ and e↓, one obtains the same scattering intensity from every atom (dashed plus solid lines). Nevertheless, there will be no complete destructive interference between the bundles coming from two neighboring atoms. This would be possible only if the two bundles were coherently superimposed. This does not, however, occur because the part of the scattered wave originating from the e↑ half of the incident wave is not coherent with that originating from the e↓ half, since the incident beam already is an incoherent superposition of e↑ and e↓. The diffraction pattern is therefore the same whether one works with polarized or with unpolarized electrons. Half-order maxima occur in both cases.

7.1.2 Electron-Molecule Scattering

Whereas the quantitative description of the polarization effects in electron diffraction from solids is quite complicated, the experimental results obtained with molecular targets are easier to understand [7.11, 12]. It was briefly mentioned in Section 3.7 that electrons that have been scattered from molecules may be polarized (Fig. 3.37). Appreciable polarization occurs if there are atoms of fairly high atomic number (causing considerable spin-orbit coupling) within the molecule. If the molecule also contains light atoms, as in organometallic compounds or halogenated hydrocarbons, the contribution to the scattering intensity from these atoms is practically unpolarized. Then the total observed polarization P does not reach the value P_1 which is caused by the heavy atoms of the molecule alone. It is instead determined by the ratio of the scattering intensities I_1 and I_2 which come from the heavy and light parts of the molecule, respectively: According to Eq. (2.16), one has

$$P = \frac{I_1 P_1 + I_2 P_2}{I_1 + I_2}. \tag{7.6}$$

The interference terms which were taken into account in (7.3) have been omitted here since they are negligible in molecular scattering at large angles where the polarization effects arise [7.11]. If $P_2 \approx 0$ we obtain from (7.6)

$$P = \frac{P_1}{1 + I_2/I_1}. \tag{7.7}$$

Consequently, the ratio I_2/I_1 can be determined by comparing the measured value P with the polarization P_1 caused by the heavy atoms alone. To do this, one can refer to the theoretical values which are available for most atoms. These calculations are reliable when the energies are not too low, as we saw in Section 3.5.

Figure 7.5 gives an example. In scattering from $Bi(C_6H_5)_3$, only the electrons scattered from the heavy bismuth atom $(Z = 83)$ are labeled by spin polarization, whereas the electrons scattered from the phenyl groups are virtually unpolarized. It can be seen from Fig. 7.5 that the polarization measured at low electron energies is small. It is much smaller than for scattering by free Bi atoms, as can be easily seen by comparison with tabulated values for Bi. With increasing energy, the polarization tends toward the values that one obtains in scattering by free Bi atoms. Thus Fig. 7.5 shows that at low energies the fraction of the electrons scattered from the Bi atom is very small; with increasing energy this fraction continually increases, as is apparent from the increasing polarization (for a more accurate analysis, see [7.11]).

This is another example of the method of "labeling" electrons by their polarization. The fact that some of the scattered electrons are labeled by their spin orientation simplifies the analysis of electron-molecule scattering by giving information about the proportion of the scattering intensities which come from the various parts of the molecule.

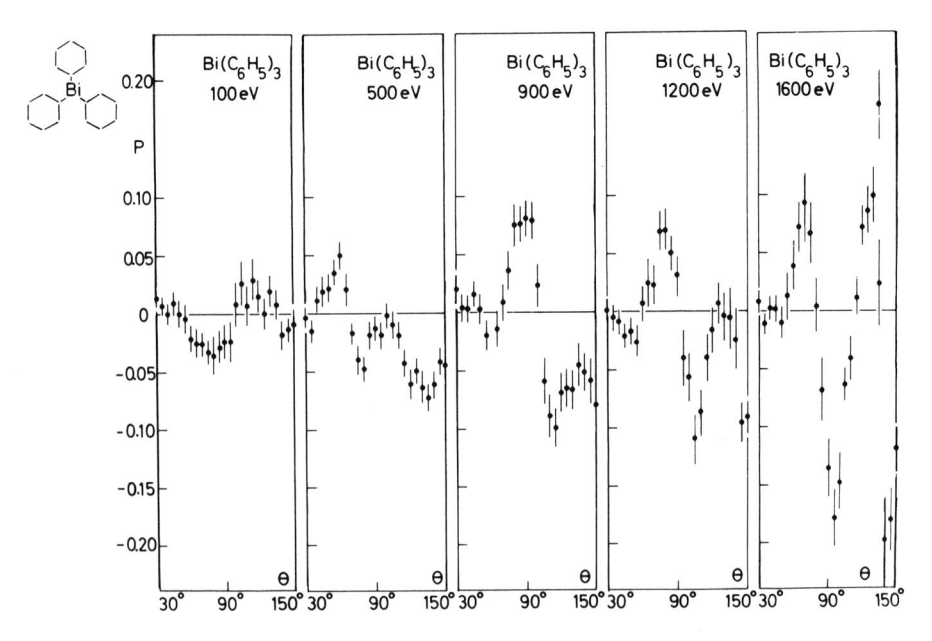

Fig. 7.5. Polarization $P(\theta)$ for scattering of slow electrons by $Bi(C_6H_5)_3$ at different energies [7.11]

The calculation of the polarization curves is less complicated for molecules than for crystals, since the influence of chemical bonding (modification of the atomic scattering potentials!), interference terms, and intramolecular plural scattering on the polarization is small. This is because, at the relatively large angles of more than 30° where the polarization is appreciable, these influences virtually disappear. Discrepancies between the theoretical values and experimental results occur merely at pronounced peaks of some polarization curves, e.g., with I_2 and Sb_4 [7.13].

7.1.3 Electron Microscopy

The previous discussion of the applications of spin polarization in electron diffraction and electron scattering suggests the idea that polarization effects might also be utilized in electron microscopy. In fact, electron microscopists were among the physicists who made the first quantitative studies of electron polarization. It is, however, hard to see any practical applications of the polarization effects in conjunction with the conventional methods of electron microscopy.

It is true that the image contrast in transmission electron microscopy comes from the basically spin-dependent electron scattering in the object; but the spin dependence of the scattering is of no significance here: In order to keep the effects of lens aberrations small, one must work with aperture angles of 0.1–1°. Electrons that are scattered in the object at larger angles do not pass through the apertures and thus do not contribute to the image intensity. Because of the small scattering angles of the imaging electrons, the Sherman function $S(\theta)$ is virtually zero in the angular range relevant to conventional transmission electron microscopy (see Sect. 3.5). Thus the electrons which contribute to the image in the conventional electron microscope are almost entirely unpolarized. For the same reason, the left-right asymmetry would be negligible if polarized electrons were used in such an electron microscope. Any polarization effects which could be expected are substantially below 10^{-6}.

The spin-dependence of the exchange scattering in magnetized materials which was discussed in Subsection 7.1.1 likewise cannot be used in transmission electron microscopy. Since one works with fast electrons and small scattering angles, no observable effects can be expected.

In some techniques recently introduced in electron microscopy, a practical application of spin-dependent interactions may be conceivable, though it will be difficult. One could, for example, think of utilizing spin-dependent Møller scattering (see Sect. 4.4) from magnetized objects in scanning transmission electron microscopy. The ratio of the cross sections for parallel and antiparallel spin directions in electron-electron scattering has a minimum near $\theta = 45°$ for the energies used in the electron microscope, so that one would have to collect the electrons scattered at those angles. Since the energy is shared between the two colliding electrons, energy analysis would allow discrimination against electrons that are elastically scattered from the atoms. The discrimination could also be achieved by detecting the two scattered electrons in coincidence. The effort going into these experiments, requiring by necessity the use of a polarized-electron source, would be great, but the returns would be small: In an iron sample magnetized to saturation, 8% of the electrons are oriented so that the observed effects would be only a few percent.

It seems then that the electron polarization microscope will not be a product of the next decade, but for the time being, must remain a product of science fiction!

7.1.4 Why Isn't Nature Ambidextrous?

Compounds containing carbon atoms bonded to four atoms or groups no two of which are alike are capable of existing in two optically active forms. They are distinguished by being respectively left- and right-handed (L- and D-enantiomorphs). Such molecular species are found in the terrestrial biosphere in only one of the two possible forms. One finds, for example, only L-amino acids in proteins and only D-sugars in carbohydrates and nucleic acids. The question of the origin of this dissymmetry has intrigued scientists since the time of Pasteur.

Among the possible causes which have been suggested are: Active seeds of organic compounds that reached the Earth from the universe, enantiomorphous minerals (e.g., left- or right-handed quartz crystals) that catalyzed or adsorbed organic compounds in a stereoselective way, or the Earth's magnetic and electric fields that influenced chemical reactions. Another explanation would be possible if an excess of either left- or right-circularly polarized light could be detected on Earth or had existed for some time during chemical evolution, since it has been shown that optical activity can be induced through stereoselective decomposition of racemic substances by circularly polarized light. (Racemic substances are mixtures of equal amounts of L- and D-enantiomorphs).

There was not much progress in the development of these and other hypotheses [7.14] until the discovery of parity violation in weak interactions 20 years ago provided a stimulus for further discussion, and molecular dissymmetry was hypothesized to be a result of the dissymmetry of interactions within the nucleus. The transfer of the dissymmetry from the nuclei to the molecules might be achieved by the longitudinally polarized electrons emitted in β decay or by circularly polarized bremsstrahlung produced by these electrons. This idea can be examined by irradiating racemic mixtures or optically inactive substances with β particles and analyzing any stereoselective degradation or synthesis that might occur. For a long time such experiments were controversial. Positive results that had been claimed by a few authors could not be confirmed by other groups [7.14a].

It seems, however, that reliable results have recently been obtained [7.15]. A racemic mixture of D- and L-leucine has been partially degraded by irradiation with a beam of longitudinally polarized 130-keV electrons. The electron polarization has been produced by scattering of slow electrons on a mercury target, as described in Chapter 3. The polarized electrons with energies in the 100-eV region were then accelerated to 130 keV and either sent into a Mott detector for polarization analysis or fired on the leucine target, after conversion of their transverse polarization into longitudinal polarization.

The polarized electron source which provided a varying polarization between 10 and 23% did not come up to the present state of the art. Nevertheless, the results are clearly beyond the statistical error limits. After irradiation times that produced 50–75% degradation of the leucine sample, the enantiomorphous composition of the undecomposed leucine was analyzed. It turned out that electrons of negative helicity (spin antiparallel to momentum), as they are emitted in β decay, bring about more extensive degradation of D-leucine than of L-leucine. Reversal of the direction of polarization showed that electrons with positive helicity engender the asymmetric decomposition of D,L-leucine in strictly the opposite sense. The extent of the asymmetry in degradation was in the 1% range in the present experiment.

The mechanism by which the asymmetric degradation is caused is not yet known; one does not even know whether it is brought about by the longitudinally polarized electrons themselves or by their bremsstrahlung. We are presently at the stage where we begin to replace speculation with facts. There is much more work to be done before we have enough reliable facts on which to base explanations.

7.1.5 High-Energy Physics

The use of polarized electrons also opens up new possibilities in high-energy physics. When describing electron scattering in the GeV region in which the large electron accelerators operate, the atomic nucleus can no longer be conceived as a point charge. Its finite size becomes significant, and the distribution of the electric charge density within the nucleus can be determined from electron scattering experiments. At these energies, the interaction between the magnetic electron moment and the nuclear moment also plays an important role in the scattering process. One of the consequences is that the cross section depends on the mutual orientation of the electron and nuclear spins. If one scatters polarized electrons by polarized nuclei, one obtains different scattering intensities according to whether the spins of the collision partners are parallel or antiparallel to each other. In analogy to what was said for atoms in Chapters 3 and 4, this difference reveals details on the structure of the nucleus which cannot be found from cross-section measurements alone.

Quite a few suggestions for high-energy experiments with polarized electrons have already been made. We will not, however, give a survey here of the numerous possibilities and suggestions that have not yet been realized in practice (some are described in the review article by DOMBEY [7.16]), but rather we will take as an example an experiment that is being carried out at the Stanford linear accelerator and briefly outline its basic idea [7.17, 18].

The experiment is designed to study electron scattering by protons with high momentum transfer. This deep inelastic scattering provides particularly interesting information about the structure of the proton and its electromagnetic interaction at high energies. From the various proton models that exist, the results obtained from scattering of unpolarized electrons favor the parton models: The electrons appear to be scattered by point-like constituents within the proton.

The purpose of the considered experiment is to investigate the spin-dependence of deep inelastic scattering of longitudinally polarized electrons by polarized protons. The main reason for using longitudinally polarized electrons is that, at high energies, transverse polarization components no longer contribute very much to the spin-dependent effects. The goal of the experiment is to measure the relative difference between the differential inelastic cross sections for parallel and antiparallel mutual orientations of the electron and proton spins. Such an asymmetry in deep inelastic scattering was first predicted by BJORKEN [7.19].

Theoretical results for the asymmetry which were obtained on the basis of various models differ significantly. Parton models based on different assumptions about the point charges within the proton (their number, spin, mass, etc.; one of the most widely known is the quark model) predict that the cross section for antiparallel spins of electron and proton should be larger than for parallel spins. The exact values of the asymmetry, however, depend on the particular model used. The aforementioned experiment is expected to facilitate a decision on the validity of the various models. Measurements of the spin-averaged cross sections which are obtained with unpolarized particles would not be very helpful since these cross sections depend only weakly on the models mentioned above.

From the numerous experimental problems that arise with high energies we want to mention only one which is of interest in connection with our general theme: How does one determine the polarization of the electrons in the GeV region? One cannot simply assume that the polarization of the electrons is the same after acceleration as it was at the time of injection into the accelerator, since it is generally affected by the electromagnetic acceleration fields. Even though calculations show that no significant depolarization is to be expected for most accelerators, a measurement of the polarization of the high-energy electrons is indispensable. Mott scattering is unsuitable for this purpose because, at these energies, it no longer produces enough asymmetry. In the present experiment Møller scattering is used (see Sect. 4.4) [7.20]. One may also use the relative differences between the cross sections for elastic scattering of polarized electrons by polarized protons with parallel and antiparallel spins. If, by suitable choice of energy and scattering angle, one works in the region where the asymmetry is largely independent of the still existing uncertainties of the proton form factors, then exact enough measurements of the electron polarization can be carried out.

High-energy experiments with polarized electrons do not necessarily have to be made with polarized targets. Possible contributions to high-energy electron scattering from weak, parity-violating interactions, for example, can be studied by using polarized electrons and unpolarized protons.

Now that the first high-energy accelerator of polarized electrons is operational, there will be an increasing interest in this area of polarized-electron physics.

7.2 $g-2$ Experiments for Measuring the Anomalous Magnetic Moment of the Electron. Electron Maser

Spin-precession experiments with polarized electrons have yielded the most accurate determination yet made of the anomalous magnetic moment of free electrons. $g-2$ has also been measured very accurately using resonance transitions between Landau levels with different spin orientations. The selection of Landau levels with a specific spin orientation by means of inhomogeneous magnetic fields can be utilized to measure $g-2$ or to build a tunable maser in the millimeter wavelength range.

One of the most impressive experiments that has been made so far with polarized electrons is the precision measurement of the magnetic moment of free electrons [7.21].

According to the Dirac theory (Sect. 3.1), the magnetic moment μ of the electron is $-\mu_B$, where $\mu_B = eh/2mc$. Hence the g factor which is defined as

$$g = \frac{-\mu/\mu_B}{s} \tag{7.8}$$

(s = spin quantum number = $\tfrac{1}{2}$) has the value 2. By observation of the shift of the fine-structure levels $2S_{1/2}$ and $2P_{1/2}$ of hydrogen and deuterium (Lamb shift), evidence was found in 1947 that the g factor differed slightly from 2 and that the Dirac theory was not completely satisfactory. This was one of the reasons for the development of quantum electrodynamics which predicts for the g factor the value

$$g = 2(1 + a) \tag{7.9}$$

with the anomaly[2]

$$a = 0.5(\alpha/\pi) - 0.32848(\alpha/\pi)^2 + (1.49 \pm 0.2)(\alpha/\pi)^3$$
$$= (1\,159\,655.3 \pm 2.5)\cdot 10^{-9} \tag{7.10}$$

(α = fine-structure constant). A precision measurement of the g factor enables quantum electrodynamics to be tested. Of particular interest are the (very small) higher terms in the expansion of a in terms of α/π, since slight deficiencies of the theory should first become noticeable there. These terms would be the first to indicate when further refinement of the theory were necessary. It is therefore desirable to make as precise a measurement as possible of the g factor.

The most accurate measurement that has been carried out to date [7.22] is based on a fact that was discussed in Subsection 3.6.3: If a polarized electron beam circulates in a homogeneous magnetic field as shown in Fig. 3.33b, the polarization alternates periodically between being longitudinal and transverse. As we have seen, the reason for this is that, since $|\mu| > \mu_B$, the precession frequency of the spins

$$\omega_p = \frac{eB}{mc}\left(\frac{1}{\gamma} + a\right) \tag{3.93}$$

---

[2] For a survey of latest theoretical developments cf. [7.21a].

is slightly greater than the cyclotron frequency

$$\omega_c = \frac{eB}{m\gamma c}. \tag{3.92}$$

Accordingly, the change between longitudinal and transverse polarization is determined by the difference frequency

$$\omega_D = \omega_p - \omega_c = a\frac{eB}{mc}. \tag{7.11}$$

By measuring the frequency ω_D, one can therefore determine the g-factor anomaly a, if the magnetic field B is known. A great advantage for the accuracy of the measurement is that it is not the total g factor that is directly measured but rather the small anomaly a.

The principle of the experiment is shown in Fig. 7.6. A short pulse ($\approx 10^{-7}$ s) of polarized electrons is injected into the magnetic field. Simultaneously a timer is switched on. The electrons circulate for a certain time T in the magnetic field, their polarization alternating periodically between longitudinal and transverse. After the storage time T which is measured by the timer, the electrons are injected into a Mott detector to analyze their polarization. The pulse height in the Mott detector depends on the state of the polarization. For transverse polarization, that

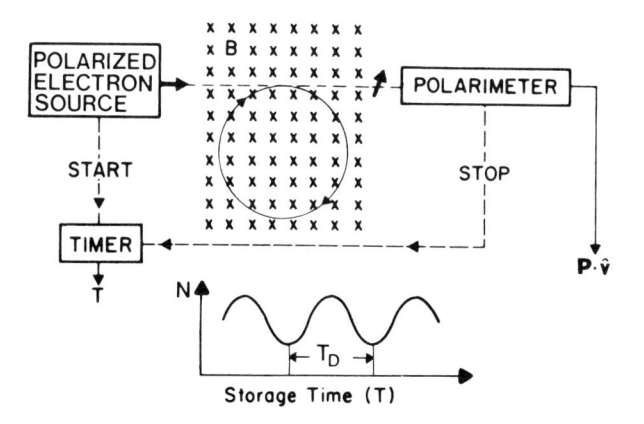

Fig. 7.6. Schematic diagram of precession experiment with polarized electrons for the measurement of $g - 2$

is $\boldsymbol{P}\cdot\boldsymbol{v} = 0$, there is a maximum or minimum of the scattering intensity according to whether $\boldsymbol{P}$ is parallel or antiparallel to the normal of the scattering plane. This scattering asymmetry was discussed in Subsection 3.3.1 [see (3.70)]. For longitudinal polarization one finds the mean value of the scattering intensity. If the procedure just described is repeated with many different storage times T one obtains a sinusoidal curve (Fig. 7.6) for N, the number of scattered electrons observed in the Mott polarimeter:

$$N \propto I(\theta)\{1 + S(\theta)P \sin \varphi_D\},$$

where φ_D is the angle between $\boldsymbol{P}$ and the velocity $\boldsymbol{v}$. Since $\varphi_D = \omega_D T + \varphi_0$ [the angle $\varphi_D(T = 0) = \varphi_0$ depends on the kind of polarized-electron source used], the observation of a complete period of this sine curve yields the desired difference frequency

$$\omega_D = \frac{2\pi}{T_D}, \tag{7.12}$$

where T_D (see Fig. 7.6) is the time in which the polarization vector makes one complete rotation more than the velocity vector. By determining the time T_D, one obtains, with (7.11), the anomaly a.

Figure 7.7 gives a general idea of the actual experimental setup. The polarized electrons were produced by scattering from a gold foil. The electron energy was in the 100-keV region and the polarization was approximately 20%. Instead of selecting the electrons scattered exactly at 90°, those which still had a small forward component of the momentum were sent through an aperture; they spiralled towards the detector with a pitch of approximately 1°. As the number of rotations of the polarization vector increases, the error in the measurement of T_D decreases because one has an increasing number of periods to use in determining T_D. It is therefore important to keep the electrons in the magnetic field as long as possible. This was achieved with the help of the cylindrical electrodes shown in Fig. 7.7. As the electrons drift across the gap between the cylinders, a momentary retarding voltage applied to the cylinders causes the electrons to lose sufficient axial velocity so that they can no longer spiral out of the magnetic field. The electrons are trapped because the magnetic field is slightly inhomogeneous and thus has field components that prevent the electrons from leaving the field ("magnetic bottle"). In this manner, stable helical paths are formed in the area denoted as the trapping region.

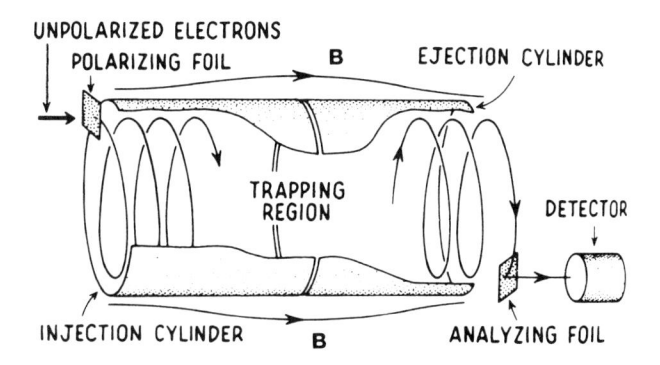

Fig. 7.7. Precession experiment with polarized electrons for the measurement of $g-2$ [7.22]

After a few milliseconds (which corresponds to several million rotations and several thousand periods T_D), a momentary acceleration voltage between the cylinders gives the electrons sufficient axial velocity to leave the magnetic field and to reach the analyzing foil of the Mott detector.

For a precision measurement of the anomaly a it does not suffice to determine T_D as accurately as possible. Equation (7.11) shows that the magnitude and local variation of the magnetic field must also be known accurately. Furthermore, the observed frequency is affected by stray electric fields which may arise from contact potentials, static charges, or the space charge of the beam. Although these stray fields are small, they determine the reliability of the last decimal places of the experimental value

$$a = (1\,159\,657.7 \pm 3.5) \cdot 10^{-9}.$$

The accuracy in the anomaly a of 3 ppm corresponds to an accuracy in the g factor, which is almost a thousand times larger, of approximately $4 \cdot 10^{-9}$. This accuracy is only possible because one does not measure the total g factor as in conventional spin-resonance methods but instead measures the small deviation $g-2$ directly.

The experimental value given above can also be written in the form

$$a = 0.5(\alpha/\pi) - 0.32848(\alpha/\pi)^2 + (1.68 \pm 0.33)(\alpha/\pi)^3,$$

where the last numerical factor in particular is affected by the exact value of the fine-structure constant α. Comparison with (7.10) shows that quantum electrodynamics makes an extremely accurate prediction. This is the most accurate comparison between theory and experiment which has been made in physics!

The successful technique discussed here has also been applied to other elementary particles. At CERN, for example, polarized muons of a few GeV have been injected into a storage ring in order to measure the g factor of free muons with the same method.

Precision measurements of the anomalous electron moment are also made with microwave techniques. Although these experiments have as yet not quite reached the accuracy of the measurement described above, the underlying principle is interesting. We shall therefore outline the basic idea which was put forward by BLOCH [7.23].

The energy of a free electron moving in a homogeneous magnetic field $\boldsymbol{B} = B\hat{e}_z$ is given in the non-relativistic limit by

$$E = \frac{p_z^2}{2m} + \left(n + \frac{1}{2}\right)\hbar\omega_0 \pm \mu B, \tag{7.13}$$

where ω_0 is the cyclotron frequency (3.92) for $\gamma = 1$. The first term in this equation corresponds to the free electron motion in the z direction, the second term corresponds to the cyclotron motion in the magnetic field (this can be considered to be composed of harmonic oscillations of frequency ω_0 in two mutually perpendicular directions); the last term is the energy for the two possible orientations of the magnetic moment $\boldsymbol{\mu}$ in the magnetic field. A rigorous derivation of relation (7.13) can be found in LANDAU and LIFSHITZ [7.24].

With the use of (7.8), (7.9), and (3.92) one obtains from (7.13) the energy terms

$$E - \frac{p_z^2}{2m} = \left(n + \frac{1}{2}\right)\hbar\omega_0 \mp \frac{1}{2}g\mu_B B = \left(n + \frac{1}{2} \mp \frac{1+a}{2}\right)\hbar\omega_0$$

which are shown in Fig. 7.8 for the lowest values of n. With suitable microwave frequencies, transitions of the electrons can be induced between the energy levels shown. With the frequency ω_0 one obtains transitions $\Delta n = \pm 1$ between different cyclotron levels; the frequency $(1 + a)\omega_0$ causes a spin flip in the same cyclotron orbit ($\Delta n = 0$), and the frequency $a\omega_0$ causes both a spinflip and a change of the cyclotron orbit. The anomaly a can be determined by measuring these frequencies, e.g., by evaluating the ratio $(1 + a)\omega_0/a\omega_0$.

Transitions between the various levels can be observed only if the levels have different populations. Even in high magnetic fields the distances

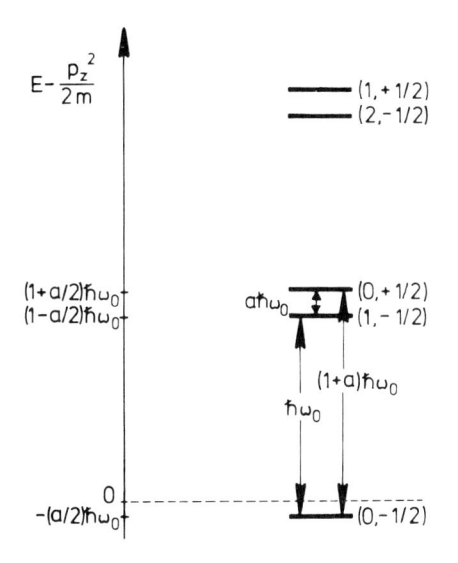

Fig. 7.8. The lowest Landau levels for an electron (not drawn to scale)

between the levels are very small in comparison with the spread of the energy distribution that free electrons usually have ($\mu_B B \approx 6 \cdot 10^{-5}$ eV at 10^4 G). Accordingly, the population differences of the various levels are generally very small. Large population differences can, however, be obtained if polarized electrons are used, since then the spin-up and spin-down states ($+\frac{1}{2}$ and $-\frac{1}{2}$ in Fig. 7.8) are differently populated.

We shall now explain with an example how the idea outlined here can be realized in practice. In the experiment shown in Fig. 7.9 an electric quadrupole field is superimposed on the magnetic field so that the electrons can be confined both radially and axially [7.25]. A hot cathode outside this trap generates a pulsed electron beam which passes through the trap and ionizes the residual gas molecules therein, so that slow electrons are obtained in the storage region.

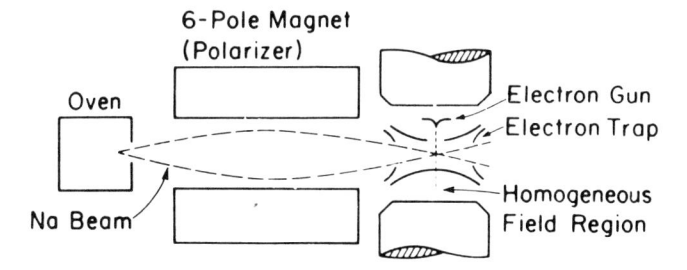

Fig. 7.9. Polarized-electron resonance experiment for measurement of $g - 2$ [7.25]

By exchange collisions with polarized sodium atoms, which also pass through the trap, the trapped free electrons become polarized in the direction of the magnetic field. This means that the spin-up and spin-down Landau levels attain different populations so that one can induce transitions between them by irradiating with the spin-flip frequencies. These transitions diminish the population differences, thus causing a drop in the electron polarization. This decrease of the polarization is utilized to indicate when spin-flip processes occur, i.e., when a suitable frequency is being used.

To monitor the decrease in the polarization, the authors exploited the fact that the cross section for inelastic scattering of polarized electrons by polarized Na atoms depends on the mutual spin orientation of the collision partners. This was discussed in detail in Subsection 4.3.1. From the relations derived there, one obtains the cross section Q' for excitation of a ^2P state:

$$Q' = \frac{1}{4}Q^s + \frac{3}{4}Q^t - \frac{1}{4}(Q^s - Q^t)P_e P_A \tag{7.14}$$

(see Problem 7.2). Q^s and Q^t are the excitation cross sections, integrated over the solid angle, for the antisymmetric (singlet) and symmetric (triplet) spin states of the two colliding electrons; P_e and P_A are the respective polarizations of the electrons and the atoms.

In the experiment discussed, excitation of the $3\ ^2P$ state is the main cause of the energy loss of the electrons. The magnitude of the corresponding cross section thus determines how fast the electrons lose their energy due to collisions with the Na atoms. Therefore, according to (7.14), the energy-loss rate of the electrons depends on their polarization. Consequently, monitoring the electron energy distribution provides a means of detecting the decrease in P_e induced by spin-flip transitions. The specific method used is to lower the trap voltage by a certain amount. Then the electrons that can overcome the potential barrier will escape. Measurement of the number of electrons remaining in the trap yields the desired information on the polarization. Finally the trap is cleared by a negative voltage pulse before another pulse from the electron gun starts the next measuring cycle.

Determination of the anomaly a from the ratio of the frequencies $(1 + a)\omega_0$ and $a\omega_0$ would be possible only if there were no electric field superimposed on the magnetic field. In actual fact, one must take into account the influence of the electric quadrupole field used to trap the electrons, which is ignored in our formulae. This correction can easily be made, since the frequency shift caused by this field can be calculated theoretically and tested experimentally.

The accuracy of the value obtained for a is determined by the precision with which the resonance frequencies can be measured, that is, by the width of the resonance lines. The present uncertainty is 260 ppm.

Similar resonance experiments have been made for some time by DEHMELT [7.26] and coworkers who pioneered work in this field and who have carried out several versions of these experiments. In one of their latest variations, the resonance transitions were detected by measuring temperature changes of the electron gas that are caused by the transitions [7.27]. The temperature was monitored by measuring the noise voltage which is induced by the motion of the electrons. The anomaly a of the g factor was determined by this method with an uncertainty of 21 ppm. The accuracy was limited by the electric field which is caused by the space charge in the trap. Since the authors have subsequently succeeded in isolating single electrons in the trap [7.28], this error source can be eliminated so that an increase in accuracy is to be expected.

Suitable conditions for resonance experiments can also be obtained by selecting electrons in some of the Landau states and thereby producing population differences. This can be done with the aid of inhomogeneous magnetic fields. The lowest state $(0, -\frac{1}{2})$ in Fig. 7.8 is seen to be paramagnetic: its energy decreases in the magnetic field. Accordingly, electrons in this state are drawn toward regions of high field strength in an inhomogeneous magnetic field. The electrons in all the other states are drawn into regions of low field strength since they behave diamagnetically (increase of energy in the field).

Consequently, a magnetic bottle such as that indicated (in a different connection) in Fig. 7.7 can be used for selecting diamagnetic Landau states: Electrons in these states are reflected from both ends into the middle and thus remain in the bottle. Electrons in the paramagnetic ground state are drawn toward the ends and are lost at the chamber walls. We must, however, keep in mind that a change in the magnetic field of 2 kG corresponds to a change in the energy of the lowest diamagnetic state of about 10^{-5} eV (for the paramagnetic ground state the energy change is only $\approx 10^{-8}$ eV). Consequently, a magnetic field with a gradient of 2 kG can only prevent an electron in this state from escaping if the electron's maximum kinetic energy in the axial direction lies below 10^{-5} eV. For higher excited states this limit is correspondingly higher.

Apart from making possible the measurement of the electron magnetic moment by resonance transitions to the depopulated paramagnetic level, the technique discussed also opens up other possibilities: By depopulating the ground state in the magnetic bottle as just described, one might generate a population inversion which would make maser operation possible. Stimulated rf transitions could then produce a gain in rf energy. One thus would have a maser in the 50 GHz range which could easily be tuned by changing the level distances through variation of the magnetic field.

A group at Stanford [7.29] has succeeded in selecting the paramagnetic ground state as follows: An electron source is located in an inhomogeneous magnetic field that decreases in the direction in which the electrons emerge. From a maximum value of 6 kG near the electron source, the field changes to a homogeneous 4-kG region, approximately 1 m long, which is used for time-of-flight measurements. Electrons in the paramagnetic ground state are decelerated as they pass through the inhomogeneous region of the magnetic field, while electrons in the diamagnetic higher states are accelerated. The energy loss of the ground-state electrons due to the deceleration is only 10^{-8} eV since the magnetic field decreases by approximately 2 kG. The electrons in the higher states gain at least 10^{-5} eV due to the acceleration. Electrons of 10^{-5} eV need approximately 0.5 ms to pass through the 1-m drift region. All those electrons which take considerably longer time must therefore be in the ground state. Consequently, ground-state electrons can be identified by their time of flight. To select polarized electrons for further experiments one must use only those electrons which are still in the drift region more than 1 ms after the electron pulse has started from the cathode.

In this experiment one can, of course, separate only those ground-state electrons whose thermal energy is not large enough to make them cross the drift region in less than 1 ms. Since the mean thermal energy of the electrons used was approximately 0.5 eV, only very few electrons (≈ 1 electron/pulse) at the extreme lower end of the thermal energy distribution fulfilled this condition.

Without going further into the numerous difficulties of such an experiment, it should be mentioned that the suppression of stray electric fields also represents a considerable problem. Contact-potential differences were suppressed so far that they had no appreciable effect on the axial motion of the 10^{-8}-eV electrons. The authors actually succeeded in detecting electrons in the ground state.

These examples show that one can make very interesting, though difficult, experiments with polarized electrons in Landau levels. Since several groups are concentrating on this problem, further progress is to be expected in the near future.

Problem 7.2: In [7.25], the total cross section for excitation of the 3 ^2P state by collisions of polarized electrons with polarized Na atoms is given in the form (7.14). Deduce this expression from the results in Subsection 4.3.1.

Solution: According to (4.31) and (4.37), the total cross section for the case in which at least one of the colliding beams is unpolarized is

$$Q = \frac{1}{4}Q^s + \frac{3}{4}Q^t, \tag{7.15}$$

where $Q^s = \sum_{i=-1}^{+1} |F_i + G_i|^2$ and $Q^t = \sum_{i=-1}^{+1} |F_i - G_i|^2$ are the cross sections for the singlet and triplet states [this can be seen immediately if one writes (4.31) in the form of (4.14)]. On the other hand, the cross section for scattering of totally polarized electrons by totally polarized atoms with the same spin direction is, according to (4.30), given by Q^t. For opposing spin directions, the cross section is, according to (4.28) and (4.29), given by $\frac{1}{2}(Q^s + Q^t)$ (because $|f|^2 + |g|^2 = \frac{1}{2}|f+g|^2 + \frac{1}{2}|f-g|^2$, as can immediately be seen by multiplying out the complex quantities).

With these results we easily find the cross section for scattering of a partially polarized electron beam by a partially polarized atomic beam (respective polarizations P_e and P_A). After separation of the partially polarized beams into totally polarized and unpolarized fractions in the respective ratios $|P_e|/(1 - |P_e|)$ and $|P_A|/(1 - |P_A|)$, we obtain

$$Q' = (1 - |P_e|)Q + |P_e|(1 - |P_A|)Q + \left\{ \begin{array}{c} P_e P_A Q^t \\ |P_A P_e| \frac{1}{2}(Q^s + Q^t) \end{array} \right\}.$$

The first term on the right-hand side describes the scattering of the unpolarized electrons by the total target; the second term is the scattering of the polarized electrons by the unpolarized part of the target; the third term is the scattering of the polarized electrons by the polarized part of the target (upper term for parallel spins, lower for antiparallel spins). Using (7.15) we obtain

$$Q' = Q + \left\{ \begin{array}{c} (Q^t - Q)P_e P_A \\ (\frac{1}{2}Q^s + \frac{1}{2}Q^t - Q)|P_e P_A| \end{array} \right\} = \frac{1}{4}Q^s + \frac{3}{4}Q^t - \frac{1}{4}(Q^s - Q^t)P_e P_A.$$

7.3 Sources of Polarized Electrons

Some of the processes discussed in the present monograph have been used to build polarized-electron sources. A comparison of the utility of these sources is made.

The discussions throughout this book have shown that an efficient source of polarized electrons would be an attractive device to have for novel investigations in various fields of physics. Indeed, there have been many attempts to utilize as polarized-electron sources some of the processes we have described. Frequently workers in the field tend to give an overoptimistic prognosis for the efficiency, as a source, of the particular process they are studying or proposing. The data that are really attained sometimes lie orders of magnitude below these predictions, and in several cases it has become obvious that simple practical limitations prevented a significant increase of the efficiency obtained.

We shall now give a brief account of the state of the art of polarized-electron sources (see also [7.30–32]). In order to be able to compare the various sources with each other, one needs common criteria of performance. To establish such criteria may at first sight appear trivial since a source is better, the higher its polarization and current are and the better its beam is collimated. One does not, however, get much further with such general ideas when one has, for example, to make the following simple decision: Would one rather have a source that yields a totally polarized beam with a moderate current or one with ten times as much current and a polarization of, say, 20%.

To answer this question, we start with the fact that in experiments with polarized-electron beams the information is mostly drawn from the relative difference or "asymmetry" A of the counting rates obtained with opposing polarizations[3] P and $-P$. The asymmetry $A = PS$ is determined by the polarization P and the quantity S which describes the spin dependence of the process to be studied. If the polarization of the electrons emitted by the source is known, the error of the desired quantity S is determined by the error of the asymmetry measurement:

$$\Delta S = \frac{\Delta A}{P}.$$

By referring to Problem 3.8, one can immediately see that the statistical error is

$$\Delta S \approx \frac{1}{\sqrt{P^2 N}}. \tag{7.16}$$

(In Problem 3.8, S was the known quantity and P had to be measured. The situation is reversed here so that P and S are also interchanged in the final result). Since the observed number of particles N, under otherwise identical conditions, is proportional to the incident current I, it follows from (7.16) that the error will decrease as $P^2 I$ increases. This quantity is therefore often taken to be a measure of the quality of a source of polarized electrons. In the example above, if the polarization is five times smaller, one needs twenty-five times as much current in order to obtain the same error limits in the same time.

There is, however, no point in using $P^2 I$ in every case as a figure of merit, even if, for example, the polarization tends to zero. The polarization must be large enough so that the spin-dependent asymmetries one wants to study do not become completely masked by asymmetries of instrumental origin. Polarizations below a few percent are of no interest for most purposes. Another case in which it would be inappropriate to use $P^2 I$ as a figure of merit is high-energy electron scattering on polarized targets (see Subsect. 7.1.5) where the electron bombardment reduces the

---

[3] This includes measurements of left-right asymmetry, if the direction of P is referred to the direction of $\hat{n}$.

polarization of the target. In such an experiment it would be pointless to compensate for small polarizations by using high intensities. When using P^2I as a somewhat rough figure of merit in the following, we should bear such restrictions in mind. If two sources have the same value of P^2I, the source with the higher polarization is usually to be preferred.

One also needs to know whether a source yields a well-collimated beam which can easily be handled by electron optical devices, that is, can be sent through lenses, filters, or spectrometers without much loss of intensity. This can be suitably described by the brightness b which is conventionally used to describe normal electron sources. It is defined as the current density per unit solid angle. A source that concentrates a high current density into a small solid angle has high brightness. If r_0 is the radius of a beam-cross-section minimum (for example, at the exit of the source) and α_0 the corresponding semi-aperture of the beam, one has the brightness (see Problem 7.3)

$$b_0 = \frac{I}{\pi^2 r_0^2 \alpha_0^2} \tag{7.17}$$

if α_0 is not too large (I = beam current).

One should also take into account that the angular divergence of an electron beam is reduced if the beam is accelerated. This is obvious because of the increase of the longitudinal momentum components during acceleration and is quantitatively described by Lagrange's law

$$r_0 \alpha_0 \sqrt{E_0} = r_f \alpha_f \sqrt{E_f}. \tag{7.18}$$

(The indices refer to the states before and after acceleration). Consequently, if two sources have the same brightness b_0 but different energies, the source with the lower energy is superior: After acceleration of the electrons to the energy E_f at which the experiment is to be carried out, this source will yield the larger brightness, since

$$b_f = \frac{I}{\pi^2 r_f^2 \alpha_f^2} = \frac{I E_f}{\pi^2 r_0^2 \alpha_0^2 E_0} = b_0 \frac{E_f}{E_0}.$$

Thus for a given brightness b_0 of the source and a given energy E_f at which the experiment is to be carried out, the brightness b_f is inversely proportional to the energy of the electrons leaving the source.

When one considers that for polarized electrons it is not I but rather P^2I that is an adequate figure of merit for the source, and that E_f is not a property of the source but rather a parameter of the experiment, it seems, from the previous discussion, sensible to use the quantity

$$q = \frac{P^2 I}{r_0^2 \alpha_0^2 E_0} \tag{7.19}$$

for comparison of the beam quality of various polarized-electron sources. q takes into account the polarization, intensity, and collimation of the beam.

In sources that utilize strong magnetic fields, the off-axis trajectories become skewed as the electrons pass through the inhomogeneous field region on their way to the field-free region where the polarized beam is to be used. This deteriorates the quality of the beam [7.33]. Comparison of different sources should therefore be made under comparable conditions, for instance, in regions free of magnetic fields.

Apart from the general criteria mentioned so far, there are further properties of polarized-electron sources that have particular relevance in some experiments. In investigations with slow electrons, for example, one often requires the energy spread of the incident beam to be small. Whether a source with a continuous or pulsed, transversely or longitudinally polarized beam is more suitable also depends on the particular experiment. These properties are not considered in the following comparison because their relevance varies from experiment to experiment.

We shall now consider the different methods for producing polarized electrons with respect to their performance as sources of polarized electrons. We shall proceed in the order in which the methods were discussed in this book and only select those cases where development of the method as a usable source of polarized electrons has actually been tried.

According to Chapter 3, electron scattering from unpolarized targets with high atomic numbers yields considerable polarization at energies from a few electron volts up to the MeV region. Numerous combinations of energy and scattering angle yield nearly total polarization. However, the intensities that can be obtained are moderate because the polarization maxima lie near cross-section minima and are moreover very narrow when they approach 1, as has been explained in Sections 3.4 and 3.5.

Polarization (or Sherman function) diagrams such as those given in Section 3.5 do not provide the best survey of the regions in which scattering yields the most favorable values. Since the scattered current I is proportional to the differential cross section $\sigma(\theta)$, diagrams of the quantity $P^2\sigma(\theta)$ or of the quantity $P^2(\theta, E)\sigma(\theta, E)\sqrt{E}$ are more suitable. The latter expression takes into account that, as the electron energy decreases, it becomes increasingly difficult to produce a high-intensity primary beam because the current of a space-charge limited electron gun is proportional to $E^{3/2}$. Since the scattered current I is directly proportional to the primary current and the quality of the source is, according to (7.19), inversely proportional to E ($E_0 = E$ = energy of electrons leaving the source), it follows that the quality of the source based on scattering is governed by the quantity

$$\frac{P^2\sigma E^{3/2}}{E} = P^2\sigma\sqrt{E}.$$

Figure 7.10 or the comparison of Fig. 7.11 with Fig. 3.21 shows that the criteria used here give a picture that is quite different from that obtained when only the polarization values are considered.

Table 7.1 gives a comparison of the various sources of polarized electrons. Only a qualitative statement is made on the values of q from (7.19), because in some cases no information on the emittance $r_0\alpha_0$ was available (cf. Problem 7.4). The value of P^2I given in the table for scattering from unpolarized targets has been obtained to within the same order of magnitude in three different laboratories (Mainz, Stanford, Münster). This indicates that a considerable improvement in this value is hardly possible.

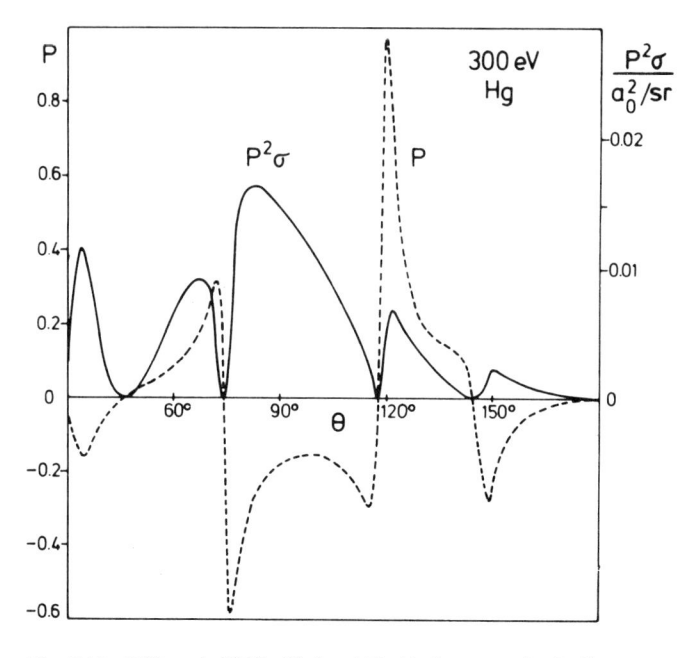

Fig. 7.10. $P(\theta)$ and $P^2(\theta)\sigma(\theta)$ for 300-eV electrons elastically scattered by Hg (a_0 = Bohr radius)

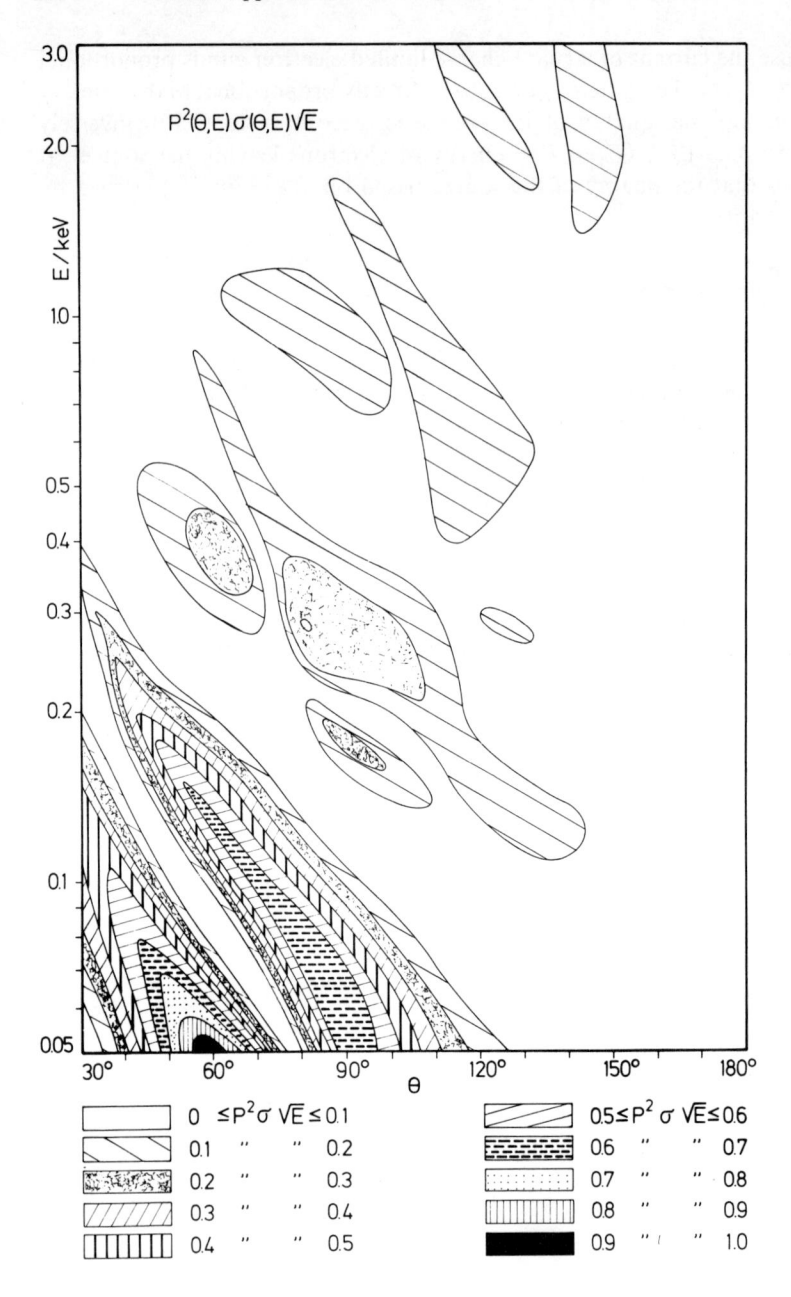

Fig. 7.11. Contours of $P^2(\theta,E)\sigma(\theta,E)\sqrt{E}$ = const. in units of $a_0^2\sqrt{\mathrm{eV}}$/sr for elastic scattering by Hg

Table 7.1. Comparison of various sources of polarized electrons

Method	P	I		P^2I (Ampere)	Beam quality q Eq. (7.19)
		d.c. (Ampere)	pulsed($\approx 1\mu s$) (el./pulse)		
Scattering from unpolarized targets	0.2	$3.5 \cdot 10^{-8}$		10^{-9}	Medium
Exchange scattering from polarized atoms	0.5 0.2		10^4 10^7	$2 \cdot 10^{-14}$ $7 \cdot 10^{-12}$	Low to medium
Photoionization of polarized atoms[a]	0.76		$8 \cdot 10^8$	10^{-8}	Medium
Fano effect	0.65 0.9	$1.5 \cdot 10^{-9}$	$3 \cdot 10^9$	$6 \cdot 10^{-10}$ $2 \cdot 10^{-11}$	Medium
Collisional ionization in optically pumped He discharge	0.3	$5 \cdot 10^{-7}$		$5 \cdot 10^{-8}$	Medium to high
Field emission from ferromagnets	0.89	10^{-6}		$8 \cdot 10^{-7}$	High

[a] This source is being constantly improved. With a somewhat reduced polarization, more than $2 \cdot 10^9$ electrons/pulse have now been obtained (contributed paper to 1975 Lepton-Photon Symposium at Stanford University).

Exchange scattering of slow electrons by polarized atoms, as discussed in Chapter 4, has also been used several times [7.34–36] to produce beams of polarized electrons. This method is particularly effective if the electrons are trapped for a while in the region containing the polarized atoms so that they have plenty of opportunity for exchange collisions. In this way their polarization gradually builds up. A typical experimental setup [7.34] works as follows.

A hot cathode emits a pulse of slow electrons of a few eV. The electrons are stored in a trap which is a combination of an electric potential well and a magnetic field. The trapped electrons collide with polarized potassium atoms which are produced using a six-pole magnet and which flow continuously through the electron trap. At the end of the trapping period of approximately 20 ms the electrons are extracted along the direction of the magnetic field. Thus 1-μs pulses of 10^4 longitudinally polarized electrons with polarizations of up to 50% were obtained (up to 10^5 electrons per pulse for smaller P). Table 7.1 compares this technique, which is described in detail in [7.36], with other methods. The second entry refers to more recent results attained at DESY using polarized hydrogen atoms as the target [7.35].

Exchange scattering may also be used to build a continuous source of polarized electrons by scattering a beam of slow electrons by a polarized atomic beam. Equation (4.24) shows that the polarization P'_e of the scattered electrons approaches the value P_A of the atoms if $|f(\theta)|^2/\sigma(\theta)$ approaches zero, that is, if virtually only exchange scattering takes place. Since the exchange processes have appreciable cross sections only at low electron energies, this technique—contrary to scattering by spinless targets—is restricted to slow primary electrons. The values of P'_e/P_A which can (in theory) be obtained have been calculated as a function of scattering angle and electron energy for rubidium between 0 and 7 eV [7.37]. For certain ranges of these parameters (e.g., $E \approx 0.03$ eV, $\theta > 90°$, and $E \approx 2$ eV, $\theta \approx 100°$) one obtains $P'_e/P_A > 0.8$, that is, the attainable electron polarization amounts to more than 80% of the polarization of the rubidium beam. Similar results have been obtained for the other alkali atoms [7.38]. Several theoretical papers on exchange scattering give as their motivation the general interest in intense sources of polarized electrons. Although such calculations are important and should be encouraged, this motivation is not convincing: a brief glance at Table 7.1 shows that other methods can provide more successful approaches to this goal.

Much work has gone into the development of polarized-electron sources from the photoionization of polarized atoms (see [5.2,3]). This was the first method used to tackle the problem of building a source. Figure 7.12 shows the scheme of an apparatus with which good results have been obtained. UV light is reflected into a longitudinally polarized atomic beam that is produced by a six-pole magnet. The photoionization takes place in a magnetic field that is strong enough to decouple electron and nuclear spins from each other (cf. Sect. 5.1). The magnitude of the photoionization cross section and that of the hyperfine interaction between electrons and nuclei both play an important role in the choice of the atoms to be used. In the latest version of the experiment lithium was used because of

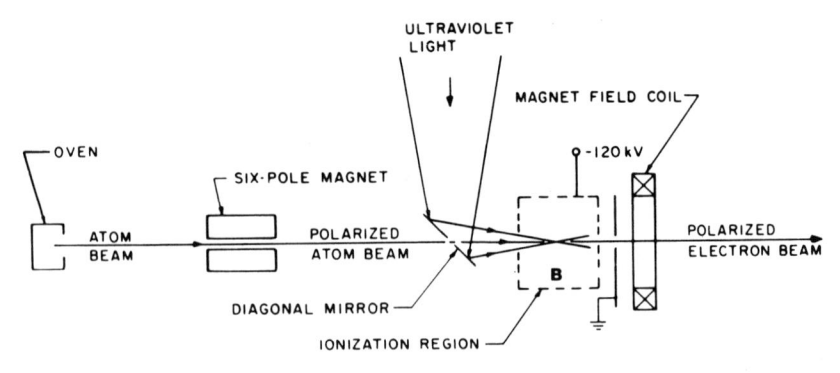

Fig. 7.12. Polarized electrons from photoionization of polarized alkali atoms [7.39]

all the alkalis it has the highest photoionization cross section. The experiment was carried out with Li^6 because this isotope has a smaller hyperfine interaction than Li^7. Lithium also has the advantage that the spin-orbit coupling which, according to Section 5.2, can drastically change the initial polarization during the photoionization process is negligibly small. In addition, one need not be afraid of an appreciable reduction of the polarization due to Li_2 molecules because their photoionization cross section is—contrary to what was said in Subsection 5.2.2 for Cs—merely of the same order of magnitude as that of the Li atoms.

The highest intensities were obtained by using a pulsed light source, since pulsed light sources such as sparks have a peak radiance in the uv which is many orders of magnitude larger than that of continuous lamps. Accordingly, the source is suitable for operation with a pulsed high-energy accelerator (see Subsect. 7.1.5). Its characteristics are presently [7.20, 40] $8 \cdot 10^8$ longitudinally polarized electrons per pulse at a polarization of 76% and a repetition rate of 180 pulses/s. The source is, however, still being improved.

A glance at the original papers cited above shows that the method is considerably more complicated in practice than would appear from the schematic diagram in Fig. 7.12. The discovery of the Fano effect (see Sect. 5.2) opened up the possibility of obtaining polarized electrons with somewhat less effort by photoionizing *un*polarized atoms. Such sources have been developed in various laboratories. The table gives as an example the data obtained by groups at Yale [7.41] and Bonn [7.42]. It should, however, be noted that the technique for producing polarized electrons which is based on the relatively new discovery of the Fano effect has yet to be optimized as a source. Better results are to be expected in the near future [7.43].

Polarized-electron production processes which have not been developed at all as polarized-electron sources are not included in Table 7.1. This applies, for example, to collisional ionization of polarized metastable atoms as discussed in Subsection 5.5.1. Attempts have, however, been made to optimize collisional ionization in an optically pumped discharge as a source of polarized electrons. We have seen in Subsection 5.5.2 that this method yields favorable results particularly in the afterglow of the discharge. This has been utilized to develop an efficient polarized-electron source [7.44].

The discharge was maintained in flowing helium, and electrons were extracted downstream, beyond the region of active discharge. The production rate of the electrons in this afterglow region was increased by injection of a reactant gas. Molecular gases like N_2 and CO_2 were particularly suitable as reactant gases because they have large cross sections both for chemi-ionization[4] and for rotational and vibrational excitation by low-energy electrons. The chemi-ionization reactions raised the electron-production rate by about a factor of 100, and the rapid thermalization of the electrons due to rotational and vibrational excitation of the molecules reduced the energy spread of the extracted beam to values ≤ 0.5 eV. The table shows that this source provides a remarkably high current of polarized electrons.

Good values can also be obtained with field emission from ferromagnetic EuS films at low temperatures. As was described in detail in Subsection 6.1.2, the method yielded a polarization of 89% under certain operational conditions. From the comparison in Table 7.1, this technique appears to be the most favorable. However, the continuous operation presumed there has not yet been achieved because of insufficient cooling of the magnetizing coils in the apparatus previously described. Nevertheless, it should be easily possible to overcome this problem by using superconducting coils. It must also be noted that the technique of field emission from ferromagnetic EuS films is not easy to perform: It requires low temperatures, ultrahigh vacuum, and a lot of know-how in producing the EuS-coated field-emission tips. In addition, the polarization attained is very sensitive to the annealing conditions maintained during the production of the tips.

On the other hand, this source does not suffer from the aforementioned (see page 198) deterioration of the beam quality caused by a magnetic field: Since the electrons emerging from the sharp field-emission tip are produced very close to the axis, skewing of the off-axis trajectories by the magnetic field is negligible. This influence does, however, affect the beam quality in photoemission from ferromagnetic materials. In that method one has either to cut down the current of the source by using only a small area near the center of the photocathode or to be content with a rather large emittance $r\alpha$ [7.45]. This problem does not occur in photoemission from nonmagnetic materials induced by circularly polarized light (see Sect. 6.2), which does not require a magnetic field for producing the polarization; it should therefore be an attractive polarized-electron source for experiments in field-free regions.

Summarizing, one can say that many successful approaches to the problem of building an intense source of polarized electrons have been made. It should, however, be pointed out that each of the sources mentioned took several years (sometimes a decade) for its development. A group new to this field may need a comparable amount of time to attain reliable performance of a source with characteristics as good as or better than those given in Table 7.1. Although sometimes it seems an author is

---

[4] Ionization processes, like $He(2^3S_1) + AB \rightarrow He(1^1S_0) + A + B^+ + e^-$, in which the chemical composition of the reactants is changed.

convinced that he has proposed the all-purpose source, a more objective point of view reveals the following picture: Many of the existing sources of polarized electrons have certain areas of application where they are superior (e.g., for studies under the conditions of ultrahigh vacuum, for scattering on gaseous beams where ultrahigh vacuum can hardly be maintained, for pulsed or for continuous operation). The simple gun of polarized electrons that is easy to operate for everybody, not just for the expert, that is applicable in any experiment, and that yields a highly polarized beam with a current which is not orders of magnitude below that of ordinary electron guns is still to be invented (in practice, not on paper).

Problem 7.3: Show that the brightness (= current density/solid angle) is given by the expression (7.17), if the aperture of the electron beam is not too large.

Solution:

$$b = \frac{i}{\Omega} = \frac{I}{\pi r_0^2 \cdot 2\pi \int_0^{\alpha_0} \sin\alpha\, d\alpha} = \frac{I}{\pi^2 r_0^2 \alpha_0^2}, \quad \text{if} \quad \sin\alpha \approx \alpha.$$

Problem 7.4: For the polarized-electron sources that are listed below with their typical emittance values, compare the numerical values of the quantity q [defined by (7.19)].

Source	$r_0\alpha_0$ (rad cm)	at energy E_0 (eV)
Scattering from unpolarized targets	$2\cdot 10^{-2}$	$3\cdot 10^2$
Photoionization of polarized atoms	10^{-2}	$7\cdot 10^4$
Fano effect (first value in Table 7.1)	$3\cdot 10^{-3}$	$2\cdot 10^3$
Collisional ionization in optically pumped discharge	$<2\cdot 10^{-2}$	$5\cdot 10^2$

Solution: In the order given above one obtains the following rounded off values:

$$q = 10^{-8}, 2\cdot 10^{-9}, 3\cdot 10^{-8}, >2.5\cdot 10^{-7}\ A/\text{rad}^2\text{cm}^2\text{eV}.$$

These numbers must not be considered as an absolute measure for the comparison of the sources because in certain experiments properties which have not been taken into account in q may be important (for example, whether the source is pulsed or continuous).

7.4 Prospects

A survey of our present knowledge about the physics of polarized electrons is given along with an assessment of the next steps that should be taken.

We conclude our treatment of the physics of polarized electrons with a few remarks on the state of the art and an evaluation of the present knowledge in this relatively new field. Pointing out the problems that should be tackled next will hopefully stimulate research leading to further understanding of the interactions of the electron spin.

The spin polarization arising from elastic electron scattering by unpolarized targets is quite well understood in the whole energy range between about 100 eV and a few MeV. The theoretical knowledge is well developed, and despite severe discrepancies between theory and experiment in the early stages of these studies, the increasing knowledge of how to avoid experimental pitfalls has finally led to complete confirmation of the theoretical curves. Only a few details concerning the exact position and magnitude of the polarization peaks seem still worth studying. It should, however, be emphasized that a series of triple scattering experiments would be necessary in order to find all the parameters that govern the scattering process. Because of the reliability of the theoretical results on the elastic cross section and the Sherman function there is confidence that the other scattering parameters are also properly described by theory. This has prevented a strong effort to study the behavior of the polarization in triple scattering experiments. The confidence may be justified, but surprises in fields which had seemed well understood have frequently furthered progress in physics.

At small energies the agreement between theory and experiment breaks down, as discussed in Subsection 3.5.2. This is a restriction of the entry "good" in the survey given in Fig. 7.13. Further work is necessary not only for a better understanding of the scattering process as such, but also for the interpretation of LEED experiments: The data obtained in LEED can be quantitatively understood only if the basic process of scattering by a single atom is clarified.

This remark is likewise true for inelastic scattering which also plays an important role in LEED. Investigations of polarization effects in inelastic scattering are only at the beginning. Though they are more difficult, theoretically and experimentally, than studies of elastic scattering, the first results obtained are encouraging (see Sect. 3.7). But it will take some time before our knowledge of this area reaches the same level as for the elastic case.

The role of spin polarization in resonance scattering is, apart from the experiment mentioned in Section 3.7, an open field. This lack of knowledge hampers the interpretation of polarization effects in other areas, such as exchange scattering, in which resonance effects may play a role.

Exchange scattering is one of the numerous examples of the fact that the spin orientation can be used for labeling electrons. Though several studies of exchange scattering by means of polarized particles exist, we are far from a comprehensive quantitative understanding of this field.

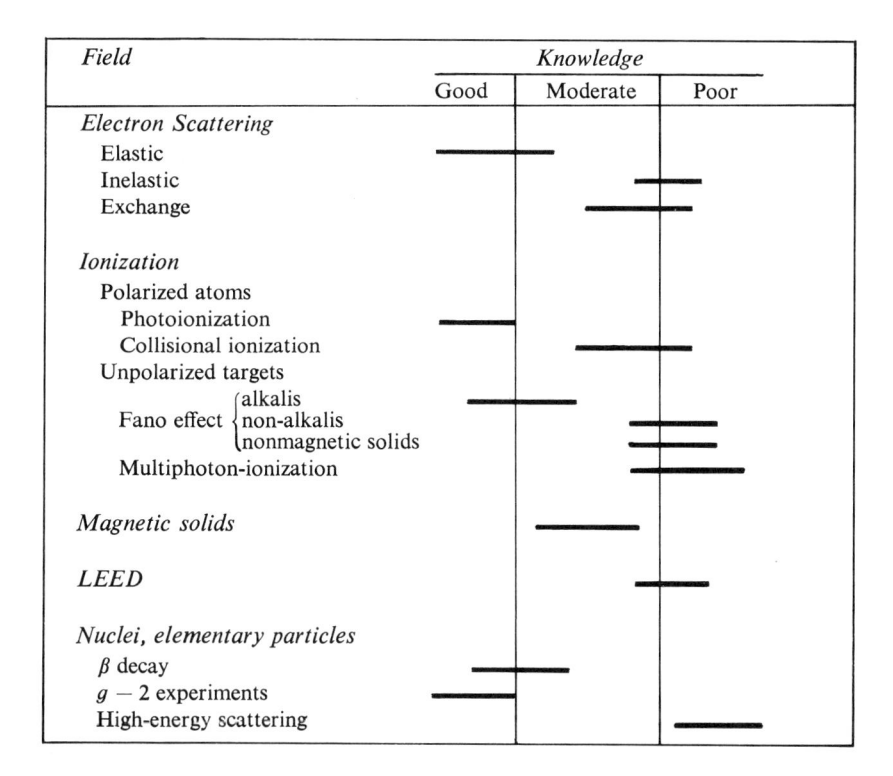

Field	Knowledge		
	Good	Moderate	Poor
Electron Scattering			
Elastic	———————		
Inelastic		———————	
Exchange		———————	
Ionization			
Polarized atoms			
Photoionization	———————		
Collisional ionization		———————	
Unpolarized targets			
Fano effect ⎰ alkalis	———————		
⎨ non-alkalis			———————
⎱ nonmagnetic solids			———————
Multiphoton-ionization			———————
Magnetic solids		———————	
LEED			———————
Nuclei, elementary particles			
β decay	———————		
g − 2 experiments	———————		
High-energy scattering			———————

Fig. 7.13. Present knowledge on electron polarization in various fields of physics

Experiments with polarized electrons of low energies and small energy spread are difficult. Accordingly, the "perfect" scattering experiment which reveals all the attainable information (also the rotation of the spin polarization by scattering) will not come about in the next few years. Exchange scattering is one of the areas that would benefit considerably from the development of more efficient sources of polarized electrons. This applies in particular to the experiments that would be necessary for a complete determination of the many parameters (discussed in Sect. 4.3) that govern inelastic scattering processes.

The generation of polarized electrons by photoionization of polarized atoms is one of the areas in which present knowledge can be rated as good. It is, however, worth noting that the discovery of the Fano effect has cast new light also on this process: The polarization of the photoelectrons may be quite different from that of the atoms owing to a significant number of spin-flip processes during photoionization at certain wavelengths. The commonly used argument that photoionization causes no spin flip since it is brought about by the electric vector of the light wave is too simple. With this argument one would not find polarization effects in electron-atom scattering either, since it is caused by the "electric" field of the atom!

The polarized electrons originating from polarized atoms can also be used as a diagnostic tool for the analysis of collision processes. Researchers are only beginning to make use of this appealing technique. Numerous such processes are conceivable; their exploitation will yield detailed knowledge on particle collisions—knowledge which so far has been obscured in the averaged results obtained with unpolarized collision partners.

Many interesting results are to be expected from studies of the polarization of photoelectrons from unpolarized targets. The only quantity that is well known now is the polarization, averaged over the solid angle, of photoelectrons produced from alkali atoms by circularly polarized light. Measurements of this quantity on substances other than alkalis are only beginning, and the possibilities for a detailed analysis of the photoionization process, including autoionizing transitions, are largely unrealized. Experimental evidence about the angular dependence of the photoelectron polarization is completely missing, and the polarization of photoelectrons produced from unpolarized atoms by unpolarized or linearly polarized light is so far not more than a theoretical prediction [5.11]. The potential of spin-polarization studies for the determination of the energy-band splitting due to spin-orbit interaction has just been recognized (see Sect. 6.2); its exploitation has not yet begun. Studies of this kind are within easy reach of present techniques and do not present the great experimental difficulties that one encounters in some areas discussed in this section, for example with problems of inelastic exchange scattering.

Another new topic of polarized-electron physics has been brought up by the advent of the laser which enables us to study multiphoton ionization. Many possibilities have been suggested, others are easy to see. Now is the time for quantitative experimental studies.

The effectiveness of polarization studies for revealing the structure of magnetic materials has been recognized and utilized for several years. The novel results obtained so far will be a stimulus for more research in the wide field of magnetism. By application of even more elaborate methods—such as the use of electron spectrometers for studying the polarization as a function of the photoelectron energy—resolution of details of the electronic structure of magnetic materials would become possible. The need to develop new models for d-band ferromagnetism and the possibility—demonstrated by field-emission experiments—of constructing spin filters that are not based on the conventional methods also call for further work.

7.4 Prospects

In LEED experiments with polarized electrons the first encouraging results are available. In conjunction with theories as they are now being developed such investigations add a new dimension to surface physics; small wonder that several groups have recently started work in this area. Further advances can be expected in the near future.

The utility of polarized-electron studies in nuclear physics has been demonstrated by the famous experiments on parity violation in weak interactions. Much experimental and theoretical work on electron polarization in β decay has been done in the past 20 years and has been treated in comprehensive monographs; thus, we have not felt it necessary to deal with this topic.

Measurements of the g factor of the electron are certainly among the greatest successes of polarized-electron physics. Nevertheless, further improvement of the high accuracy obtained seems tempting in order to search for limits of the present theory of electromagnetic interactions.

Investigations using polarized electrons have just been started in high-energy electron scattering. Such experiments may contribute to the surprises to be expected in this field.

Only a few of the many aspects of polarized-electron physics have been studied for any length of time; elastic electron scattering is such a case. Some topics, such as polarized electrons from nonmagnetic solids, were not even discussed a few years ago. Though all the areas treated in this book are within reach of present experimental techniques, progress in some areas, such as polarization effects in inelastic scattering, depends crucially on major advances in the methods for producing and analyzing electron polarization.

References

Complete reference lists are to be found in the review articles and monographs given below. The primary sources given are either directly referred to in the text or have appeared later than the review papers listed

Chapter 1

1.1 N. F. MOTT, H. S. W. MASSEY: *The Theory of Atomic Collisions* (Clarendon Press, Oxford 1965) Chapt. IX
1.2 P. S. FARAGO: Advan. Electronics Electron Phys. **21**, 1 (1965)

Chapter 2

2.1 N. F. MOTT, H. S. W. MASSEY: *The Theory of Atomic Collisions* (Clarendon Press, Oxford 1965) Chapt. IX, Sect. 5
2.2 H. A. TOLHOEK: Rev. Mod. Phys. **28**, 277 (1956)
2.3 U. FANO: Rev. Mod. Phys. **29**, 74 (1957)

Chapter 3

3.1 N. F. MOTT, H. S. W. MASSEY: *The Theory of Atomic Collisions* (Clarendon Press, Oxford 1965) Chapt. IX
3.2 D. M. FRADKIN, R. H. GOOD: Rev. Mod. Phys. **33**, 343 (1961)
3.3 L. I. SCHIFF: *Quantum Mechanics*, 3rd ed. (McGraw Hill Book Co., New York 1968) Chapt. 13
3.4 H.A. BETHE, E. E. SALPETER: In *Handbuch der Physik* (Springer, Berlin-Göttingen-Heidelberg 1957) Vol. XXXV, p. 133 ff.
3.5 H. F. SCHOPPER: *Weak Interactions and Nuclear Beta Decay* (North-Holland Publ. Co., Amsterdam 1966);
H. FRAUENFELDER, A. ROSSI: In *Methods of Experimental Physics*, ed. by L. C. L. YUAN, C. S. WU (Academic Press, New York 1963) Vol. 5, Pt. B, p. 214;
H. FRAUENFELDER, R. M. STEFFEN: In *Alpha, Beta, and Gamma Ray Spectroscopy*, ed. by K. SIEGBAHN (North-Holland Publ. Co., Amsterdam 1968) Vol. II, p. 1431;
L. A. PAGE: Rev. Mod. Phys. **31**, 759 (1959)
3.6 J. W. MOTZ, H. OLSEN, H. W. KOCH: Rev. Mod. Phys. **36**, 881 (1964)
3.7 J. KESSLER, N. WEICHERT: Z. Physik **212**, 48 (1968); W. BÜHRING: Z. Physik **212**, 61 (1968) and references therein
3.8 H. ÜBERALL: *Electron Scattering from Complex Nuclei* (Academic Press, New York 1971) Pt. A
3.9 K. JOST, J. KESSLER: Z. Physik **195**, 1 (1966)
3.10 D. W. WALKER: Advan. Phys. **20**, 257 (1971)
3.11 W. BÜHRING: Z. Physik **208**, 286 (1968)
3.12 H. DEICHSEL, E. REICHERT, H. STEIDL: Z. Physik **189**, 212 (1966)
3.13 K. SCHACKERT: Z. Physik **213**, 316 (1968)

3.14 J. KESSLER: Rev. Mod. Phys. **41**, 1 (1969)
3.15 W. ECKSTEIN: Internal Rept. IPP 7/1, Institut für Plasmaphysik, Garching bei München 1970
3.16 G. HOLZWARTH, H. J. MEISTER: *Tables of Asymmetry, Cross Sections and Related Functions for Mott Scattering of Electrons by Screened Au and Hg Nuclei* (University of Munich 1964)
3.17 M. FINK, A. C. YATES: At. Data **1**, 385 (1970);
M. FINK, J. INGRAM: At. Data **4**, 129 (1972);
D. GREGORY, M. FINK: At. Data and Nucl. Data Tables **14**, 39 (1974)
3.18 J. VAN KLINKEN: Nucl. Phys. **75**, 161 (1966)
3.19 W. EITEL, K. JOST, J. KESSLER: Z. Physik **209**, 348 (1968)
3.20 R. J. v. DUINEN, J. W. G. AALDERS: Nucl. Phys. A **115**, 353 (1968)
3.21 N. ERNST: Diplomarbeit, Universität Münster 1973
3.22 H. DEICHSEL: Z. Physik **164**, 156 (1961)
3.23 V. BARGMANN, L. MICHEL, V. L. TELEGDI: Phys. Rev. Letters **2**, 435 (1959)
3.24 H. J. MEISTER: Z. Physik **166**, 468 (1962)
3.25 P. S. FARAGO: Advan. Electronics Electron Phys. **21**, 1 (1965)
3.26 W. EITEL, J. KESSLER: Z. Physik **241**, 355 (1971)
3.27 D. H. MADISON, W. N. SHELTON: Phys. Rev. A **7**, 514 (1973)
3.28 R. A. BONHAM: J. Electron Spectrosc. **3**, 85 (1974)
3.29 T. HEINDORFF, J. HÖFFT, E. REICHERT: J. Phys. B: Atom. Molec. Phys. **6**, 477 (1973);
T. SUZUKI, H. TANAKA, M. SAITO, H. IGAWA: J. Phys. Soc. Jap. **39**, 200 (1975)
3.30 W. FRANZEN, R. GUPTA: Phys. Rev. Letters **15**, 819 (1965)
3.31 W. HILGNER, J. KESSLER: Z. Physik **221**, 305 (1969)
3.32 D. W. WALKER: Phys. Rev. Letters **20**, 827 (1968)
3.33 A. C. YATES: Phys. Rev. Letters **20**, 829 (1968)

Chapter 4

4.1 P. S. FARAGO: Rep. Progr. Phys. **34**, 1055 (1971)
4.2 P. G. BURKE, H. M. SCHEY: Phys. Rev. **126**, 163 (1962)
4.3 J. BYRNE: J. Phys. B: Atom. Molec. Phys. **4**, 940 (1971)
4.4 J. BYRNE, P. S. FARAGO: J. Phys. B: Atom. Molec. Phys. **4**, 954 (1971)
4.5 H. G. DEHMELT: Phys. Rev. **109**, 381 (1958)
4.6 R. E. COLLINS, B. BEDERSON, M. GOLDSTEIN: Phys. Rev. A **3**, 1976 (1971)
4.7 B. BEDERSON: In *Atomic Physics* 3, ed. by S. J. SMITH, G. K. WALTERS (Plenum Press, New York 1973) p. 401, and references therein
4.8 D. HILS, M. V. MCCUSKER, H. KLEINPOPPEN, S. J. SMITH: Phys. Rev. Letters **29**, 398 (1972)
4.9 E. M. KARULE, R. K. PETERKOP: *Atomic Collisions III*, ed. by Y. IA. VELDRE (Latvian Academy of Sciences, Riga, USSR 1965). Available through SLA Translation Service, John Crerar Library, 86 East Randolph St., Chicago, IL. Translation No. TT-66-1239
4.10 P. G. BURKE, L. F. B. MITCHELL: J. Phys. B: Atom. Molec. Phys. **7**, 214 (1974)
4.11 P. S. FARAGO: J. Phys. B: Atom. Molec. Phys. **7**, L28 (1974)
4.12 D. W. WALKER: J. Phys. B: Atom. Molec. Phys. **7**, L489 (1974)
4.13 K. RUBIN, B. BEDERSON, M. GOLDSTEIN, R. E. COLLINS: Phys. Rev. **182**, 201 (1969)
4.14 B. BEDERSON: Comments Atomic Molecular Phys. **1**, 65 (1969); **2**, 160 (1971)
4.15 W. LICHTEN, S. SCHULTZ: Phys. Rev. **116**, 1132 (1959)
4.16 E. U. CONDON, G. H. SHORTLEY: *The Theory of Atomic Spectra* (Cambridge Univ. Press 1967)
4.17 L. I. SCHIFF: *Quantum Mechanics* (McGraw Hill Book Co., New York 1968) 3rd ed.
4.18 H. KLEINPOPPEN: Phys. Rev. A **3**, 2015 (1971) (serveral formulae in this paper have to be corrected)

4.19 P. S. Farago, J. S. Wykes: J. Phys. B: Atom. Molec. Phys. **2**, 747 (1969)
4.20 J. Wykes: J. Phys. B: Atom. Molec. Phys. **4**, L19 (1971)
4.21 G. F. Hanne, J. Kessler: J. Phys. B: Atom. Molec. Phys. **9**, 791 (1976)
4.22 G. F. Hanne: J. Phys. B: Atom. Molec. Phys. **9**, 805 (1976)
4.23 J. C. Steelhammer, S. Lipsky: J. Chem. Phys. **53**, 1445 (1970)
4.24 R. A. Bonham: J. Chem. Phys. **57**, 1604 (1972)
4.25 K. Blum, H. Kleinpoppen: Phys. Rev. A **9**, 1902 (1974)
4.26 A. M. Bincer: Phys. Rev. **107**, 1434 (1957)
4.27 C. Møller: Ann. Physik **14**, 531 (1932)
4.28 G. Holzwarth: Z. Physik **191**, 354 (1966)
4.29 K. Ulmer: Z. Physik **135**, 232 (1953)
4.30 A. Ashkin, L. A. Page, W. M. Woodward: Phys. Rev. **94**, 357 (1954)
See also References 7.34 to 7.38.

Chapter 5

5.1 E. Fues, H. Hellmann: Phys. Z. **31**, 465 (1930)
5.2 G. Baum, U. Koch: Nucl. Instrum. Meth. **71**, 189 (1969)
5.3 V. W. Hughes, R. L. Long, Jr., M. S. Lubell, M. Posner, W. Raith: Phys. Rev. A **5**, 195 (1972)
5.4 U. Fano: Phys. Rev. **178**, 131 (1969); Addendum: Phys. Rev. **184**, 250 (1969)
5.5 U. Heinzmann, J. Kessler, J. Lorenz: Z. Physik **240**, 42 (1970)
5.6 G. Baum, M. S. Lubell, W. Raith: Phys. Rev. A **5**, 1073 (1972)
5.7 J. C. Weisheit: Phys. Rev. A **5**, 1621 (1972)
5.8 D. W. Norcross: Phys. Rev. A **7**, 606 (1973)
5.9 V. L. Jacobs: J. Phys. B: Atom Molec. Phys. **5**, 2257 (1972)
5.10 N. A. Cherepkov: Zh. Exsp. Teor. Fiz. **65**, 933 (1973); [Sov. Phys. -JETP **38**, 463 (1974)]
5.11 C. M. Lee: Phys. Rev. A **10**, 1598 (1974)
5.12 H. A. Stewart: Phys. Rev. A **2**, 2260 (1970)
5.13 G. V. Marr, R. Heppinstall: Proc. Phys. Soc. London **87**, 293 (1966)
5.14 J. Berkowitz, W. A. Chupka: J. Chem. Phys. **45**, 1287 (1966)
5.15 U. Heinzmann, H. Heuer, J. Kessler: Phys. Rev. Letters **34**, 441 (1975)
5.16 U. Heinzmann, H. Heuer, J. Kessler: Phys. Rev. Letters **36**, 1444 (1976)
5.17 P. Lambropoulos: Phys. Rev. Letters **30**, 413 (1973)
5.18 P. S. Farago, D. W. Walker: J. Phys. B: Atom. Molec. Phys. **6**, L280 (1973)
5.19 H. D. Zeman: In *Electron and Photon Interactions with Atoms*, ed. by H. Kleinpoppen, M. R. C. McDowell (Plenum Press, New York 1976) p. 581
5.20 P. S. Farago, D. W. Walker, J. S. Wykes: J. Phys. B: Atom. Molec. Phys. **7**, 59 (1974)
5.21 P. Lambropoulos: Phys. Rev. A **9**, 1992 (1974)
5.22 P. Lambropoulos, M. Lambropoulos: In *Electron and Photon Interactions with Atoms*, ed. by H. Kleinpoppen, M. R. C. McDowell (Plenum Press, New York 1976) p. 525
5.23 B. Donnally, W. Raith, R. Becker: Phys. Rev. Letters **20**, 575 (1968)
5.24 H. A. Bethe, E. E. Salpeter: In *Handbuch der Physik* (Springer, Berlin-Göttingen-Heidelberg 1957) Vol. XXXV, p. 370ff.
5.25 M. V. McCusker, L. L. Hatfield, G. K. Walters: Phys. Rev. A **5**, 177 (1972)
5.26 J. C. Hill, L. L. Hatfield, N. D. Stockwell, G. K. Walters: Phys. Rev. A **5**, 189 (1972)
5.27 P. J. Keliher, F. B. Dunning, M. R. O'Neill, R. D. Rundel, G. K. Walters: Phys. Rev. A **11**, 1271 (1975)
5.28 B. L. Donnally, R. Faber, J. Gates, C. Volk: Bull. Am. Phys. Soc. **18**, 141 (1973)
5.29 G. F. Drukarev, V. D. Ob'edkov, R. K. Janev: Phys. Letters **42A**, 213 (1972)
5.30 V. D. Ob'edkov: ZhETF Pis. Red. **21**, 220 (1975) [JETP Letters **21**, 98 (1975)]
5.31 L. D. Schearer: Phys. Rev. A **10**, 1380 (1974)

Chapter 6

6.1 H. C. Siegmann: Phys. Reports, **17**, 37 (1975)
6.2 H. Alder, M. Campagna, H. C. Siegmann: Phys. Rev. B **8**, 2075 (1973)
6.3 G. Busch, M. Campagna, H. C. Siegmann: J. Appl. Phys. **41**, 1044 (1970)
6.4 D. T. Pierce, H. C. Siegmann: Phys. Rev. B **9**, 4035 (1974)
6.5 K. Sattler, H. C. Siegmann: Phys. Rev. Letters **29**, 1565 (1972)
6.6 G. Busch, M. Campagna, H. C. Siegmann: Phys. Rev. B **4**, 746 (1971)
6.7 M. Campagna, H. C. Siegmann: Phys. kondens. Materie **15**, 247 (1973)
6.8 P. M. Tedrow, R. Meservey: Phys. Rev. B **7**, 318 (1973)
6.9 S. F. Alvarado, W. Eib, F. Meier, D. T. Pierce, K. Sattler, H. C. Siegmann, J. P. Remeika: Phys. Rev. Letters **34**, 319 (1975)
6.10 G. Obermair: Z. Physik **217**, 91 (1968)
6.11 B. A. Politzer, P. M. Cutler: Phys. Rev. Letters **28**, 1330 (1972)
6.12 G. Chrobok, M. Hofmann, G. Regenfus: Phys. Letters **26A**, 551 (1968)
6.13 W. Gleich, G. Regenfus, R. Sizmann: Phys. Rev. Letters **27**, 1066 (1971)
6.14 N. Müller, W. Eckstein, W. Heiland, W. Zinn: Phys. Rev. Letters **29**, 1651 (1972)
6.15 U. Heinzmann, K. Jost, J. Kessler, B. Ohnemus: Z. Physik **251**, 354 (1972)
6.16 K. Koyama, H. Merz: Z. Physik B **20**, 131 (1975)
6.17 C. Kittel: *Quantum Theory of Solids* (Wiley and Sons, New York 1963) Chapt. X
6.18 K. Koyama, H. Merz: Verhandl. Deutsche Physikal. Gesellschaft **10**, 527 (1975)
6.19 D. T. Pierce, F. Meier, P. Zürcher: Phys. Letters **51A**, 465 (1975)
6.20 J. R. Chelikowsky, M. L. Cohen: Phys. Rev. Letters **32**, 674 (1974)
6.21 D. T. Pierce, F. Meier, P. Zürcher: Appl. Phys. Letters **26**, 670 (1975)
6.22 G. Regenfus, P. Sütsch: Z. Physik **266**, 319 (1974)

Chapter 7

7.1 P. J. Jennings, B. K. Sim: Surface Sci. **33**, 1 (1972)
7.2 R. Feder: Phys. Stat. Sol. (b) **62**, 135 (1974)
7.3 M. R. O'Neill, M. Kalisvaart, F. B. Dunning, G. K. Walters: Phys. Rev. Letters **34**, 1167 (1975)
7.4 C. J. Davisson, L. H. Germer: Phys. Rev. **33**, 760 (1929)
7.5 C. E. Kuyatt: Phys. Rev. B **12**, 4581 (1975)
7.6 R. Feder: Phys. Stat. Sol. (b) **58**, K137 (1973)
7.7 J. C. Slater: *The Self-consistent Field for Molecules and Solids* (McGraw Hill Book Co., New York 1974) Vol. 4
7.8 P. W. Palmberg, R. E. DeWames, L. A. Vredevoe: Phys. Rev. Letters **21**, 682 (1968)
7.9 T. Suzuki, N. Hirota, H. Tanaka, H. Watanabe: J. Phys. Soc. Japan **30**, 888 (1971)
7.10 C. Rau, R. Sizmann: Phys. Letters **43A**, 317 (1973)
7.11 W. Hilgner, J. Kessler, E. Steeb: Z. Physik **221**, 324 (1969)
7.12 A. C. Yates: Phys. Rev. **176**, 173 (1968). See also Refs. 3.31 to 3.33.
7.13 J. Kessler, J. Lorenz, H. Rempp, W. Bühring: Z. Physik **246**, 348 (1971)
7.14 W. Thiemann: Naturwiss. **61**, 476 (1974)
7.14a A. S. Garay, L. Keszthelyi, I. Demeter, P. Hrasko: Nature **250**, 332 (1974)
7.15 W. A. Bonner, M. A. van Dort, M. R. Yearian: Nature **258**, 419 (1975)
7.16 N. Dombey: Rev. Mod. Phys. **41**, 236 (1969)
7.17 G. Domokos, S. Kovesi-Domokos, E. Schonberg: Phys. Rev. D **3**, 1191 (1971)
7.18 S. M. Berman, J. Primack: Phys. Rev. D **9**, 2171 (1974); the author is also indebted to Profs. W. Raith and M. S. Lubell for private communications on this experiment
7.19 J. D. Bjorken: Phys. Rev. D **1**, 1376 (1970)

7.20 P. S. Cooper, M. J. Alguard, R. D. Ehrlich, V. W. Hughes, H. Kobayakawa, J. S. Ladish, M. S. Lubell, N. Sasao, K. P. Schüler, P. A. Souder, G. Baum, W. Raith, K. Kondo, D. H. Coward, R. H. Miller, C. Y. Prescott, D. J. Sherden, C. K. Sinclair: Phys. Rev. Letters **34**, 1589 (1975)
7.21 A. Rich, J. C. Wesley: Rev. Mod. Phys. **44**, 250 (1972)
7.21a J. M. Jauch, F. Rohrlich: *The Theory of Photons and Electrons* (Springer, New York-Heidelberg-Berlin 1976) p. 530ff.
7.22 J. C. Wesley, A. Rich: Phys. Rev. A **4**, 1341 (1971), corrected by S. Granger and G. W. Ford: Phys. Rev. Letters **28**, 1479 (1972)
7.23 F. Bloch: Physica **19**, 821 (1953)
7.24 L. D. Landau, E. M. Lifshitz: *Quantum Mechanics* (Pergamon Press, London 1965)
7.25 G. Gräff, F. G. Major, R. W. H. Roeder, G. Werth: Phys. Rev. Letters **21**, 340 (1968);
G. Gräff, E. Klempt, G. Werth: Z. Physik **222**, 201 (1969)
7.26 H. G. Dehmelt: Advan. Atom. Molec. Phys. **3**, 53 (1967); **5**, 109 (1969)
7.27 F. L. Walls, T. S. Stein: Phys. Rev. Letters **31**, 975 (1973)
7.28 D. Wineland, P. Ekstrom, H. Dehmelt: Phys. Rev. Letters **31**, 1279 (1973)
7.29 L. V. Knight: Thesis, Stanford University, 1965
7.30 K. Jost, H. D. Zeman: HEPL 590 (Stanford University, 1971)
7.31 K. Jost: In *Proceedings of the VI Yugoslav Symposium and Summer School on the Physics of Ionized Gases* (Institute of Physics, Belgrade 1972) p. 37
7.32 J. Kessler: In *Atomic Physics* 3, ed. by S. J. Smith, G. K. Walters (Plenum Press, New York 1973) p. 523
7.33 W. Raith: In *Atomic Physics*, ed. by B. Bederson, V. W. Cohen, F. M. J. Pichanick (Plenum Press, New York 1969) p. 389
7.34 D. M. Campbell, H. M. Brash, P. S. Farago: Phys. Letters **36A**, 449 (1971)
7.35 R. Krisciokaitis, W. Y. Tsai: Nucl. Instrum. Meth. **83**, 45 (1970);
R. Krisciokaitis, W. K. Peterson: In *Electronic and Atomic Collisions: Abstracts of Papers, VIII ICPEAC, Belgrade, 1973* (Institute of Physics, Belgrade 1973) p. 257
7.36 P. S. Farago: Rep. Progr. Phys. **14**, 1054 (1971)
7.37 V. D. Ob'edkov, I. Kh. Mossallami: Vestnik Leningrad University **22**, 43 (1971)
7.38 E. Karule: J. Phys. B: Atom. Molec. Phys. **5**, 2051 (1972)
7.39 R. L. Long, Jr., W. Raith, V. W. Hughes: Phys. Rev. Letters **15**, 1 (1965)
7.40 M. J. Alguard, R. D. Ehrlich, V. W. Hughes, J. S. Ladish, M. S. Lubell, K. P. Schüler, G. Baum, W. Raith: Abstracts of contributed papers, Fourth International Conference on Atomic Physics, ed. by J. Kowalski, H. G. Weber, Heidelberg 1974, p. 373
7.41 G. Baum, M. S. Lubell, W. Raith: Bull. Am. Phys. Soc. **16**, 586 (1971)
7.42 W. v. Drachenfels, U. T. Koch, R. D. Lepper, T. M. Müller, W. Paul: Z. Physik **269**, 387 (1974)
7.43 W. v. Drachenfels, U. T. Koch, T. M. Müller, H. R. Schaefer: Phys. Letters **51A**, 445 (1975)
7.44 P. J. Keliher, R. E. Gleason, G. K. Walters: Phys. Rev. A **11**, 1279 (1975)
7.45 E. Garwin, F. Meier, D. T. Pierce, K. Sattler, H. C. Siegmann: Nucl. Instr. Meth. **120**, 483 (1974)

The following papers, which have appeared while this book has been in the press, are particularly relevant to the fields indicated in parentheses.

N. Müller: Phys. Letters **54A**, 415 (1975) (Field emission)
E. Kisker, G. Baum, A. H. Mahan, W. Raith, K. Schröder: Phys. Rev. Letters **36**, 982 (1976) (Field emission)
R. Feder: Phys. Rev. Letters **36**, 598 (1976) (LEED)
E. H. A. Granneman, M. Klewer, K. J. Nygaard, M. J. Van der Wiel: J. Phys. B: Atom. Molec. Phys. **9**, L87 (1976) (Multiphoton Ionization)

Subject Index

Afterglow 152, 203
Alternating polarization 186, 187
Analogy between photoionization and electron scattering 136, 138
Analyzer, *see also* Polarization measurement 6, 59, 175
Analyzing power 59, 94
Angle between scattering planes 51
Angular dependence of photoelectron polarization 130, 133, 138, 208
Angular divergence, *see* Collimation of electron beam
Annealing 162, 204
Anomaly of magnetic moment, *see* Magnetic moment
Antiferromagnetic compounds 154, 159, 177 – 179
Antisymmetric wave function 88, 91, 115
Asymptotic form of scattering wave functions 33 – 36, 39, 40
  for Coulomb potential 36
Atom-surface collisions 152, 178
Autoionization 139 – 143, 149, 208
Azimuthal dependence of cross section 38, 39, 43

Background electrons 71, 78, 79, 137
Backscattered electrons, *see* Background electrons
Band model, *see also* Density of states 153, 158, 161
Band splitting 153, 154, 160, 163, 165 – 167, 208
β decay 58, 59, 71, 120, 182, 183, 207, 209
  longitudinal polarization in 58
Born approximation 88, 116
Bragg condition 172, 173

Bremsstrahlung by polarized electrons 86, 182, 183
Brightness 197, 205

Change of polarization by scattering 45 – 49, 92 – 94, 99, 100, 105, 110 – 114
Charge-cloud polarization 68
Chemical bonding, influence on polarization 181
Chemi-ionization 152, 203, 204
Circular polarization of emitted light 100, 102, 104 – 110
  connection of light intensity and excitation cross section 109
  for polarization analysis 111
Clebsch-Gordan coefficients 103, 126, 127
Coherent superposition of electron waves 37, 171, 172
Coherent superposition of spin states 11, 19, 28, 34, 129, 138
Coincidence experiments 108, 120, 182
Collimation of electron beam 197, 198
Collisional ionization 147 – 152, 201, 203, 205, 207, 208
Commutation relations 7, 26
Components of polarization vector 12, 15, 46, 59, 72, 73, 76, 130, 132, 133
Compton scattering 86
Conservation of angular momentum, *see also* Constant of the motion 27, 36, 126
Conservation of degree of polarization in scattering 47
Conservation of direction of polarization in scattering 47
Conservation of parity, *see* Parity
Conservation of spin 110, 151
Constant of the motion 26 – 28, 103

Construction of polarization from cross sections 54, 55, 135, 136
Coulomb field, scattering by 61–63
   validity of results for 62
Cross section, *see* Photoionization cross section, Scattering cross section
Curie temperature 153, 156, 162
Cyclotron frequency 81, 187, 190

Deep inelastic scattering 184
Definition of polarization 1, 12–14, 29, 30
   Lorentz invariant 30
Degradation, asymmetric 183
Degree of polarization 13, 17, 47, 55
   in diagonal density matrix 16, 17
Density matrix 14–20
   diagonal form of 16, 17, 19, 20
   of scattered state 41, 44
Density of states 154, 155, 158, 160
Depolarization 113, 114, 185
Diagnostic possibilities of polarization studies, *see* Information
Diamagnetic electron states 193, 194
Difference between photoelectron polarization from solids and atoms 164
Differential cross section, *see* Scattering cross section
Dimers 137, 203
Dipole-dipole interaction 87, 177
Dipole matrix elements 128, 129, 132
Dirac theory 21–32, 186
Direct cross section 89, 92–94, 97, 98, 101, 102, 104, 105
Discharge 149–152, 203
Double scattering experiments 49–51, 69–71, 96, 97
   problems of 71

Effective radius of atom 53
Eigenfunctions of angular momentum 103, 126–128, 131, 132
   of spin, *see* Spin functions
Elastic scattering, *see also* Exchange scattering, elastic 33–80, 82, 83, 85, 86, 198–201, 205–209
Electron diffraction, *see* LEED
Electron-electron scattering 86, 116–121, 181
   for polarization analysis 117, 120, 121, 185

Electronic structure of magnetic materials, *see also* Density of states 155, 158, 208
Electron maser 193
Electron microscopy 181, 182
Electron-molecule scattering, polarization in 86, 179–181
Electron trap, *see also* Magnetic bottle 188, 191, 192, 201
Electrostatic field, influence on polarization 80, 81
Emittance 199, 204, 205
Enantiomorphs 182, 183
Energy-loss rate 192
Energy spread 198, 204, 207
Exchange cross section 89, 92–94, 97–99, 101, 102, 104, 105, 113
Exchange interaction in solids 153, 154, 160
Exchange scattering 68, 87–121, 151, 175–178, 181, 192, 201, 202, 206, 207
   elastic 87–100, 116–121
   in forward direction 113
   inelastic 100–116, 192, 194, 195, 208
Excitation time 102, 114
Expectation value of spin operator 12, 29, 143

Fano effect 125–131, 145, 158, 201, 203, 205, 207
   experimental studies 136–138
   illustration of 133–136
Fermi energy 153–156, 158–161, 163
Ferrites 159
Ferromagnetism, *see* Ferromagnets
Ferromagnets 153–163, 175, 176, 204
Field emission 153, 159–163, 169, 201, 204, 208
Figure of merit for Mott detector 79, 82, 83
Figure of merit for polarized electron sources 196–198
Filter lens 70, 97
Fine-structure constant 186, 189
Fine-structure splitting 22, 102, 104, 105, 113, 114, 145
Free electrons in Dirac theory, *see* Plane wave in Dirac theory

g factor 81, 82, 185–193, 209
$g-2$ experiments 185–193, 207

Half-order maxima 178
Hamiltonian function 22
  for free particle 21
Hamiltonian operator 21, 22, 26, 128
Helicity 56, 183
High-energy scattering 183 – 185, 196, 207, 209
Hund's rule 155
Hyperfine interaction, decoupling of 97, 105, 124, 148, 202

Identical atoms 172
Identical particles 88
Illustration of change in magnitude of polarization 52 – 55
Incoherent superposition of spin states 18, 19, 34, 99, 129, 138, 139, 179
Inelastic scattering, *see also* Exchange scattering, inelastic 84, 85, 173, 175, 176, 184, 206, 207, 209
Information from polarization measurements, *see also* Labeling 47, 48, 89, 92, 93, 100, 105, 107, 108, 113, 114, 137, 143, 149, 152, 155, 166, 169, 171, 175, 184, 208
Inner potential 173, 174
Instrumental scattering asymmetry 77, 79
  determination of 77
Integral cross section 105
Interband transitions 166, 167
Interference structure of cross section 53

Junctions with superconducting films 158, 159

Klein-Gordon equation 22
  linearization of 23

Labeling by polarization 86, 114, 150, 180, 206
Lagrange's law 197
Lamb shift 186
Landau levels 190 – 194
Laser 146, 208
LEED (low-energy electron diffraction) 171 – 179, 206, 207, 209
Left-handedness, *see* Enantiomorphs
Left-right asymmetry, *see also* Scattering asymmetry 39, 45, 95, 175, 181

in double scattering experiment 51, 71
  illustration of 55, 56
  in LEED 172, 175
Level-crossing technique 148
Lifetime 100, 105
Light elements, difference to heavy elements of polarization in scattering 63, 68
Linear polarization of emitted light 106
Lock-in technique 97
Longitudinal polarization 39, 80, 82, 86, 118, 120, 184, 186, 198
  and parity conservation 56 – 58
Lorentz force 3 – 5

Magnetic bottle 188, 193
Magnetic field caused by relative motion 31, 34, 52, 81
Magnetic moment 30, 31, 82, 153, 185
  anomalous 185 – 193
Maxima of polarization 64 – 66, 85, 137, 141, 206
  connection with minima or maxima of cross section 55, 66, 85, 136, 141, 173, 176, 198
Measurement of parameters determining scattering 47, 48, 92, 100
Measurement of polarization components, *see also* Polarization measurement 13, 14
  in triple scattering experiment 73, 75
Measurement of spin components 8
Metastables 101, 147 – 152
Mirror image of experiment 56 – 59, 101
Mirror inversion, connection with parity inversion 57
Mixtures of polarized and unpolarized beams 17, 55, 92, 105, 195
Mixtures of spin states 14 – 20
Mixtures of totally polarized beams 18 – 20, 53, 60, 61
Møller scattering, *see* Electron-electron scattering
Mott detector 39, 76 – 80, 175
  calibration of 78

Mott detector
  comparison with analyzer for
    polarized light 79
  efficiency of 79, 80, 142
  figure of merit for, *see* Figure of merit
    at low energies 80
Mott scattering, *see also* Elastic
  scattering 39, 43
Multiphoton ionisation 143–146, 207, 208
  comparison with Fano effect 145
Multiple scattering 71, 77, 121, 173,
  175, 176

Néel temperature 178
Neutron scattering 177
Normal to the scattering plane 43

Optical activity 182, 183
Optical pumping 149, 150
Orbital angular momentum,
  non-commutativity with H 26
Orientation of polarization, *see also*
  Spin, orientation of 12, 19, 155, 167

Paramagnetic electron state 193, 194
Parity conservation 56, 57, 101
  consequence for polarization of
    scattered beam 56 – 58
Parity violation 58, 59, 182, 185, 209
  consequence for β decay 58
Partial polarization, *see also*
  Mixtures 1, 14 – 19
Partial waves 35, 36
  method of, in relativistic
    scattering 35 – 37
Particular solutions of Dirac
    equation 35
  azimutal dependence of 36
Parton models 184
Pauli matrices 7
  generalization of 27
Pauli principle, influence on
  scattering 86 – 88
Penning ionization 149 – 152
Perfect scattering experiment 94, 100,
  207
Phases of complex scattering
  amplitudes 48, 92, 100
Photoelectrons, polarized, *see*
  Photoemission, Photoionization
Photoemission from magnetic
  solids 153 – 159, 161, 204, 207, 208

Photoemission from nonmagnetic
  solids 163 – 169, 204, 207, 209
Photoionization by circularly polarized
  light, *see also* Fano effect 139 – 146,
  163 – 169
Photoionization cross section
  134 – 136, 140 – 143, 202
  minima of 135 – 137, 141
  resonances in 140, 141
Photoionization by linearly polarized
  light 138, 139, 208
Photoionization of polarized
  atoms 123 – 125, 201 – 203, 205, 207
Photoionization by unpolarized
  light 138, 208
Photoionization of unpolarized targets,
  see Photoionization by circularly
  polarized light
Photon-spin transfer in
  photoionization 133
Plane wave in Dirac theory 24, 25, 28,
  29
Plural scattering 71, 77, 78, 181
Polarization as consequence of
  scattering asymmetry 60, 61
Polarization, definition of, *see*
  Definition of polarization
Polarization dependence of cross
  section 40 – 44
Polarization filter, *see* Spin filter
Polarization measurement, *see also*
  Mott detector, Left-right asymmetry,
  Electron-electron scattering,
  Measurement of polarization
  components 39, 86, 111, 175, 185,
  192
Polarization by scattering, *see also*
  LEED, electron-molecule
  scattering 44, 45, 84, 85, 87,
  121, 201, 202
  connection with asymmetry in Mott
    scattering 60
Polarization after scattering of arbitrary
  beam 45, 46, 48, 49
Polarization at small scattering
  angles 64
Polarization transformer 73, 80– 82
Polarization vector 12
Polarized atoms, production of 3, 123,
  124
Polarized electrons, production of, *see*
  Sources of polarized electrons

Subject Index   221

Polarized light, production of   3
Polarized muons   190
Polarized protons   184
Polarizer   3, 6, 59, 175
Polarizing power   59, 94
Population difference of spin states   1, 191 – 193
Population inversion   193
Potential barrier in field emission   160, 162, 163
  at surface, see Surface potential barrier
Precession experiment   187 – 190
Precession frequency   81, 82, 186
Precession of polarization in electrostatic field   81
Precession of polarization in magnetic field   81, 82, 186 – 188
Precession of polarization during scattering   52
Probability of spin eigenvalues   8, 16, 127
Probing depth   159, 177
Pure spin states   9 – 14

Quality of source of polarized electrons   196, 198, 199, 201, 205
Quantum electrodynamics   186, 190
Quark model   184
Quench field   148

Racemic mixtures   182, 183
Radial matrix elements, role in Fano effect   133, 136, 137, 145
Reactant gas   203
Reactions between metastables   151, 152
Reasons for polarization studies, see also Information from polarization measurements   1, 2
Recoil atoms   92, 93, 95 – 97
Reduction of polarization component by scattering   46, 67
Relative motion, transformation of fields in   31, 34, 52, 81
Relativistic covariance of Dirac theory   23, 30
Relativistic energy law   22, 25
Relativistic generalization of Schrödinger equation   21
Reliability of knowledge about polarization effects in scattering   62, 68, 71, 206

Resonance experiments   189, 190 – 193
Resonance scattering, polarization in   84, 85, 206
Resonances in photoelectron polarization   139 – 141
Reversal of polarization in scattering   76, 111 – 113
Right-handedness, see Enantiomorphs
Rotation of polarization (spin) in electromagnetic fields, see Precession
Rotation of polarization by scattering   46, 47, 66, 67, 94, 100, 175, 207
  illustration of   52
Rotational excitation   204

Saturation of intermediate states   146
Saturation of magnetization   156
Scattering amplitudes, see also Spin-flip amplitude   37, 63, 88
  for Coulomb field   61
  for direct scattering   89, 92, 93, 101, 116
  for electron-electron scattering   116
  for exchange scattering   89, 92, 93, 101, 115, 116
  in LEED   171, 172
Scattering asymmetry, see also Left-right asymmetry, Spin-up-down asymmetry   38, 55, 95
  dependence on polarization components   44, 59
  instrumental, see Instrumental scattering asymmetry
  reduction by plural scattering   77
Scattering of basic spin states   34, 35, 94
Scattering cross section, see also Total cross section   33, 38, 39, 43, 51, 89, 90, 101, 113
  connection of extrema with polarization maxima, see Maxima of polarization
  for Coulomb field   62
  difference for e↑ and e↓   54, 55
  for direct scattering, see Direct cross section
  for electron-electron scattering   117 – 120
  for exchange scattering, see Exchange cross section

Scattering cross section
  for longitudinally polarized
    beam   42, 118, 120
  for screened Coulomb field   53, 63
  for unpolarized beams   42, 90, 101
Scattering matrix S 41
Scattering phase   37
Scattering plane   39
Schrödinger equation   21
Screening of Coulomb field   62
Selection rules   125, 139, 144, 150, 166
Selection of spin states   193, 194
Sherman function $S(\vartheta)$ 38, 181, 206
  for Coulomb field   62, 63
  effective   77 – 79
  measurement of   49, 51, 69, 70
  role in scattering   38, 43, 45, 59, 60
  for screened Coulomb field   64, 65
Singlet state   90, 114, 192, 195
  excitation of   114
Six-pole magnet   123, 124, 201
Skewing of trajectories   198, 204
Sources of polarized electrons   70, 121, 123, 125, 137, 152, 195 – 205, 207
  comparison of   201
  criteria of performance, see Figure of merit
Spin   1, 7
  eigenvalues of, see also Probability of spin eigenvalues   7, 11
  formal description of   7 – 9, 26 – 28
  orientation of   7, 8, 11, 19, 29, 155, 167
Spin components, measurement of   8
Spin-dependent forces, see also
  Spin-orbit coupling, Dipole-dipole interaction   87, 112
Spin filter   2, 3, 6, 96, 163, 208
Spin flip   34, 36, 112, 128, 134, 158, 190, 192, 207, 208
  during electron transport in solids   158, 169
Spin-flip amplitude   34, 46
Spin functions   7
  antisymmetric   90
  four-component   24, 25
  of free electrons   25, 28
  in rest frame   29
  symmetric   90
  two-component   7 – 9, 11
Spin operator   7, 26, 27, 31
  non-commutativity with H 27, 32

Spin-orbit coupling   31, 34, 80, 136, 175
  role of, in electron scattering   34, 36, 52, 53, 63, 64, 87, 99, 112, 114, 179
  role of, in photoemission from solids   164 – 167
  role of, in photoionization   128, 133 – 136, 145, 203
Spin-orbit energy   32
Spin-orbit potential as cause of polarization in scattering   53
Spin-orbit relaxation time   102, 105, 114
Spin transformer, see Polarization transformer
Spin-up-down asymmetry   95
Statistical error of polarization experiment   79, 82, 83, 196
Stern-Gerlach experiment   2 – 6
Storage time   187
Stray electric fields   189, 194
Superconducting films, see Junctions
Superposition, see Coherent, Incoherent superposition
Surface magnetism   159, 177, 178
Surface potential barrier   173
Synchrotron radiation   142

Target density, see also Thickness of analyzer foil   71, 72
Thermal energy   194
Thickness of analyzer foil   72, 77, 79
Thomas precession   32, 82
Threshold, excitation at   108
Time-of-flight measurements   194
Torque on magnetic moment   52
Total cross section   106 – 109, 192, 194, 195
Total polarization   1, 13, 17
Transverse polarization   29, 38, 42, 80, 82, 118, 120, 184, 186, 198
  as consequence of parity conservation   56, 57
Triple scattering experiment   72 – 76, 96, 112 – 114, 206
Triplet state   90, 192, 195
  excitation of   110 – 116
Tunneling   160, 162, 163
Tunneling detector   159
Tunneling experiments   158, 159
Two-photon transitions   144 – 146

Subject Index

Uncertainty principle 4, 5
Uncertainty of theoretical polarization data at low energies 68
Unit cells, magnetic and chemical 177, 178
Unpolarized electrons, *see also* Mixtures of totally polarized beams 9, 17
 illustration of polarization after scattering of 52 – 55
 scattering by polarized atoms 90, 93, 101, 105

Vacuum level 165, 166, 168
Vector model 126 – 128
Vibrational excitation 204

Wavelength dependence of polarization 136 – 138, 140 – 142, 157 – 159, 164, 167, 168
Weighting factors 14, 15
Wien filter 82
Work function, reduction of 158

Topics in Applied Physics
Founded by H. K. V. Lotsch

Vol. 2 **Laser Spectroscopy** of Atoms and Molecules
Edited by *H. Walther*
137 figures, 22 tables. XVI, 383 pages. 1976

Contents: *H. Walther*, Atomic and Molecular Spectroscopy with Lasers. – *E. D. Hinkley*, Infrared Spectroscopy with Tunable Lasers. – *K. Shimoda*, Double-Resonance Spectroscopy of Molecules by Means of Lasers. – *J. M. Cherlow* and *S. P. S. Porto*, Laser Raman Spectroscopy of Gases. – *B. Decomps, M. Dumont*, and *M. Ducloy*, Linear and Nonlinear Phenomena in Laser Optical Pumping. – *K. M. Evenson* and *F. R. Petersen*, Laser Frequency Measurements, the Speed of Light, and the Meter.

Vol. 13 **High-Resolution Laser Spectroscopy**
Edited by *K. Shimoda*
132 figures. Approx. 320 pages. 1976

Contents: *K. Shimoda*, Line Broadening and Narrowing Effects. – *B. Jacquinot*, Atomic Beam Spectroscopy. – *V. S. Letokhov*, Saturation Spectroscopy. – *J. L. Hall* and *J. A. Magyar*, High Resolution Saturated Absorption Studies of Methane and some Methyl-Halides. – *V. P. Chebotayev*, Three-Level Laser Spectroscopy. – *S. Haroche*, Quantum Beats and Time-Resolved Fluorescence Spectroscopy. – *N. Bloembergen* and *M. D. Levenson*, Doppler-Free Two-Photon Absorption Spectroscopy.

Vol. 14 **Laser Monitoring of the Atmosphere**
Edited by *E. D. Hinkley*
84 figures. Approx. 360 pages. 1976

Contents: *S. H. Melfi*, Remote Sensing for Air Quality Management. – *V. E. Zuev*, Laser-Light Transmission through the Atmosphere. – *R. T. H. Collis* and *P. B. Russell*, Lidar Measurement of Particles and Gases by Elastic Backscattering and Differential Absorption. – *H. Inaba*, Detection of Atoms and Molecules by Raman Scattering and Resonance Fluorescence. – *E. D. Hinkley, R. T. Ku*, and *P. L. Kelley*, Techniques for Detection of Molecular Pollutants by Absorption of Laser Radiation. – *R. T. Menzies*, Laser Heterodyne Detection Techniques.

Springer-Verlag Berlin Heidelberg New York

Topics in Physics Today

Beam-Foil Spectroscopy Edited by *S. Bashkin*

Contents: S. Bashkin: Introduction. – *S. Bashkin:* Instrumentation. – *I. Martinson:* Wavelengths Measurements and Level Analysis. – *L. Curtis:* Lifetime Measurements. – *I. Sellin:* Autoionizing Levels. – *H. Marrus:* Studies of H-like and He-like Ions of High Z. – *W. Whaling, L. Heroux:* Applications to Astrophysics. – *O. Sinanoglu:* Fundamental Calculation of Level Lifetimes. – *W. Wiese:* Systematic Effects in Z-Dependence of Oscillator Strengths. – *J. Macek, D. J. Burns:* Coherence, Alignment, and Orientation Phenomena.

R. Beck, W. Englisch, K. Gürs

Table of Laser Lines in Gases and Vapors

IV, 130 pages. 1976. (Springer Series in Optical Sciences, Vol. 2)

Laser Spectroscopy

Proceedings of the 2nd International Conference, Mégève, France, June 23 – 27, 1975
Edited by *S. Haroche* et al. 230 figures, 30 tables. X, 468 pages (5 pages in French). 1975. (Lecture Notes in Physics, Vol. 43)
Contents: Spectroscopy 1.– Tunable Lasers 1. – Spectroscopy 2. – Spectroscopy 3. – Tunable Lasers 2. – Laser Isotope Separation. – Spectroscopy 4. – Spectroscopy 5. – Titles and Abstracts of Post-Deadline Papers.

Surface Physics

58 figures. VI, 125 pages. 1975. (Springer Tracts in Modern Physics, Vol. 77)
Contents: *Wissmann, P.,* The Electrical Resistivity of Pure and Gas Covered Metal Films: Experimental. Structure of the Films. The Resistivity of Pure Metal Films. The Temperature Dependence of Resistivity of Pure Metal Films. Resistivity Change Due to Gas Adsorption. Resistivity and Heat of Adsorption. Concluding Remarks. – *Müller, K.,* How Much Can Auger Electrons Tell Us About Solid Surfaces?: The Distribution $N(E)$. Energy Levels of the Sample. Some Important Interactions. Instrumentation. Auger Transitions. Inspection of an Auger Spectrum. The Qualitative Element Analysis. Steps Towards Quantitative Analysis. Deconvolution. Line Shape and the Density of States. Line Shape and Chemical Environment. Auger Electrons from Compound Solids. The Concept of Cross Transitions. A Potential Example: Cs-C. Metal Oxides. Conclusion.

Springer-Verlag Berlin Heidelberg New York